Biology Living Systems

Chapter Assessment

McGraw-Hill

New York, New York   Columbus, Ohio   Mission Hills, California   Peoria, Illinois

A GLENCOE PROGRAM

Biology
Living Systems

Student Edition
Teacher Wraparound Edition
Study Guide, SE and TE
Investigating Living Systems, SE and TAE
Probing Levels of Life, SE and TAE
Chapter Assessment
Videodisc Correlations
Science and Technology Videodisc Series Teacher Guide
Transparency Package
Concept Mapping
Exploring Environmental Issues
Critical Thinking/Problem Solving
Spanish Resources
Lesson Plans
Computer Test Bank

Copyright © 1994 by Glencoe/McGraw-Hill.
All rights reserved.
Permission is granted to reproduce the material contained herein on the condition that such material be reproduced only for classroom use; be provided to students, teachers, and families without charge; and be used solely in conjunction with the **Biology: Living Systems** program. Any other reproduction, for use or sale, is prohibited without prior written permission of the publisher.

Send all inquiries to:

Glencoe/McGraw-Hill
936 Eastwind Drive
Westerville, Ohio 43081

ISBN 0-02-826301-4

Printed in the United States of America

3 4 5 6 7 8 9 10 POH 00 99 98 97 96 95

Contents

This booklet provides materials to assess your students' learning of concepts from *Biology Living Systems*. For each of the thirty chapters of the Student Edition, there are two levels of questions.

The *Understanding Concepts* section tests recall and comprehension of important facts and vocabulary presented in the chapter. Question formats include true-false, matching, completion, multiple choice, and short answer.

The *Interpreting and Applying Concepts* section requires students to use several different higher-order learning skills. For some questions, students will need to interpret data and discover relationships presented in graphs and tables. In other instances, students may be asked to read about an experiment and then apply their understanding of chapter concepts and scientific methods to analyze and explain the results. There are also essay-type questions which require students to use major concepts to solve problems, to compare and contrast situations, and to make generalizations and inferences.

Answers to all questions are provided on the reduced pages at the back of the assessment booklet.

CHAPTER 1 🐛 ASSESSMENT

BIOLOGY–THE SCIENCE OF LIFE

Understanding Concepts

Classify each organism as a producer (p), consumer (c), or decomposer (d) by writing the correct letter in the space at the left.

_____ 1. algae   _____ 5. garden snake   _____ 9. tiger

_____ 2. ant   _____ 6. goldfish   _____ 10. tree

_____ 3. bread mold   _____ 7. grass

_____ 4. cow   _____ 8. mushroom

In the space at the left, write the letter of the word or phrase that best completes the statement or answers the question.

_____ 11. Two products of cellular respiration are
- **a.** oxygen and glucose.
- **b.** oxygen and water.
- **c.** carbon dioxide and water.
- **d.** carbon dioxide and glucose.

_____ 12. Each of the following is a biologically-based solution to controlling the zebra mussel invasion *except*
- **a.** depriving them of oxygen.
- **b.** scraping them off surfaces.
- **c.** introducing one of their predators.
- **d.** preventing fertilization of their eggs.

_____ 13. In producers, chlorophyll and sunlight are necessary for the process of
- **a.** homeostasis.
- **b.** photosynthesis.
- **c.** cellular respiration.
- **d.** reproduction.

_____ 14. The closing of its shell when a clam is removed from its watery environment is an example of how the clam maintains its
- **a.** growth.   **b.** development.   **c.** evolution.   **d.** homeostasis.

_____ 15. In a food chain involving a mouse and the snake that eats it, the mouse and snake are
- **a.** both consumers.
- **b.** both producers.
- **c.** consumer and decomposer, respectively.
- **d.** producer and consumer, respectively.

Write the word or phrase that best completes the statement.

16. The supply of energy that producers need to make food comes from _____.

17. Physical traits that make an organism well suited to its environment are called _____.

18. The broadest division into which organisms may be classified is a(n) _____.

19. The increase in the amount of living material in an organism is referred to as _____.

20. The changes an organism undergoes in reaching its adult form is its _____.

Copyright © by the Glencoe Division of Macmillan/McGraw-Hill School Publishing Company   **CHAPTER 1 ASSESSMENT**   **1**

BIOLOGY–THE SCIENCE OF LIFE

Interpreting and Applying Concepts

The graphs in Figure 1–1 depict human growth in height from birth through age 18 for males and females. Study the graphs and then answer the questions.

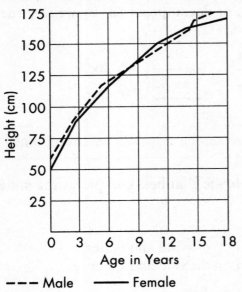 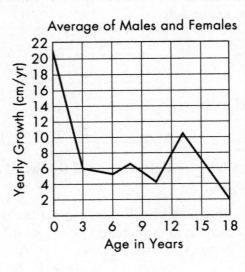

Figure 1–1

1. On the first graph, what indicates that growth takes place from birth to age 3?

   _____

   _____

   _____

2. From ages 6 to 9, about how many centimeters does a female child grow?

   _____

3. During what range of ages do females generally grow faster than males?

   _____

4. About how many centimeters taller are males at age 15 than females at the same age?

   _____

5. During which years do all children tend to grow the fastest? _____

6. Describe the trend in human growth depicted in the second graph.

   _____

   _____

   _____

 Copyright © by the Glencoe Division of Macmillan/McGraw-Hill School Publishing Company

BIOLOGY—THE SCIENCE OF LIFE

Interpreting and Applying Concepts continued

7. Complete Table 1–1 comparing cellular respiration and photosynthesis.

Table 1–1

Cellular Respiration	Photosynthesis
energy is given off	
	occurs in green plants
produces carbon dioxide and water	
	carbon dioxide and water combine

Figure 1–2 shows the number of usable energy units available at each level of a food chain. Use the figure to answer the questions.

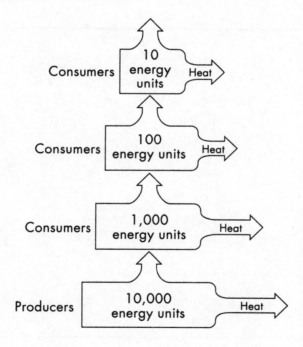

Figure 1–2

8. How many energy units are lost in the form of heat as energy flows from producer to first consumer in this food chain? What percent of the original amount does that represent?

_____

9. As you go up the food chain, how does the amount of energy available at each level compare

with the previous level? _____

_____

Copyright © by the Glencoe Division of Macmillan/McGraw-Hill School Publishing Company   CHAPTER 1 ASSESSMENT   **3**

BIOLOGY—THE SCIENCE OF LIFE

Interpreting and Applying Concepts   continued

One food chain based on this model consists of corn grain, a corn-fed chicken, and a human who eats chickens. Suppose the corn contains 100 energy units.

10. How many energy units are available to the chicken from corn? _____

11. Which is a more energy-efficient diet for humans, a diet of corn or a diet of chicken? Explain.

_____

_____

_____

12. Which would support a larger number of people with less energy input, a diet of corn or a diet of corn-fed chicken? Explain your reasoning. _____

_____

     Copyright © by the Glencoe Division of Macmillan/McGraw-Hill School Publishing Company

CHAPTER 2 &❧ ASSESSMENT

BIOLOGY AS A SCIENCE

Understanding Concepts

In the space at the left, write the letter of the phrase from Column B that best matches the term in Column A.

Column A	Column B
_____ 1. control group	**a.** applied science
_____ 2. data	**b.** information gathered from observations
_____ 3. experiment	**c.** process that produces a body of knowledge about nature
_____ 4. experimental group	
_____ 5. hypothesis	**d.** group in which all the variables remain constant
_____ 6. science	**e.** tests an explanation
_____ 7. technology	**f.** statement that explains and relates data
_____ 8. theory	**g.** hypothesis that has withstood the test of time
	h. group in which the independent variable is changed

In the space at the left, write the letter of the word or phrase that best completes the statement or answers the question.

_____ 9. A good way of showing comparisons among groups when data are not dependent on one another is a
　　a. line graph.　　**b.** pie graph.　　**c.** bar graph.

_____ 10. A three-dimensional image of an object is produced by a
　　a. transmission electron microscope.
　　b. scanning electron microscope.
　　c. compound light microscope.

_____ 11. The length times the width times the height of a bricklike cell is the cell's
　　a. solid volume.　　**b.** weight.　　**c.** area.

_____ 12. A mass of 726 grams is equal to
　　a. 7.26 kilograms.　　**b.** 0.726 kilogram.　　**c.** 72.6 kilograms.

_____ 13. The starting point of all scientific research is a(n)
　　a. controlled experiment.　　**b.** hypothesis.　　**c.** observation.

Copyright © by the Glencoe Division of Macmillan/McGraw-Hill School Publishing Company

BIOLOGY AS A SCIENCE

Interpreting and Applying Concepts

Based on her field observations, a scientist hypothesized that a certain species of insect requires a specific range of air temperatures in order to live. She tested her hypothesis by exposing the larval stage of this species to varying air temperatures. The data from her experiments are shown below. Use the data to answer the following questions.

TEMPERATURE (°C)	SURVIVAL RATE (%)
15	0
16	20
17	60
18	80
19	90
20	100
21	100
22	80
23	70
24	30
25	0

1. What is the independent variable in the experiment? _____

2. What is the dependent variable in the experiment? _____

3. What type of graph would you use to display these data? Give reasons for your choice.

_____

_____

4. Draw the graph you described in Question 3.

Copyright © by the Glencoe Division of Macmillan/McGraw-Hill School Publishing Company

BIOLOGY AS A SCIENCE

Interpreting and Applying Concepts continued

Use the graph constructed for Question 4 to answer the following questions.

5. What relationship does the graph illustrate? _____

_____

6. At what air temperature(s) did all the organisms survive? _____

7. At what air temperature(s) did all the organisms die? _____

8. Do the data support the scientist's original hypothesis? Explain. _____

_____

9. Based on the data, what range of air temperatures offers these organisms the best chance for

   survival? _____

10. Based on the data, what range of air temperatures seems to pose the greatest danger to these

    organisms? _____

11. What would you expect to happen to a group of these organisms if the air temperature was

    28°C? _____

12. In conducting her research, the scientist used heat lamps to control the air temperature
    surrounding the organisms. No variable other than temperature of air was noted. Can she be

    sure that the results are linked to air temperature? _____

_____

13. Suppose you wanted to repeat this experiment to verify the results. Explain how you would go

    about investigating the original hypothesis. _____

_____

CHAPTER 3 🐌 ASSESSMENT

MATTER AND ENERGY

Understanding Concepts

In the space at the left, write TRUE if the statement is true. If the statement is false, change the italicized word or phrase to make it true.

_____ 1. A solution in which the concentration of hydrogen ions is greater than the concentration of hydroxide ions is *a base.*

_____ 2. *An ionic bond* results when atoms combine by sharing electrons.

_____ 3. *A chemical formula* shows the number and kind of each atom in a compound.

_____ 4. DNA and RNA belong to the class of biological compounds called *lipids.*

The terms in the first pair are related to each other. In the space at the left, write the letter of the second pair of terms that are related in the same way.

	First Pair		**Second Pair**

_____ 5. atom, element
            a. electron, nucleus
            b. molecule, compound
            c. proton, energy level

_____ 6. glucose, carbohydrate
            a. hemoglobin, nucleic acid
            b. maltose, protein
            c. cholesterol, lipid

_____ 7. ammonia, base
            a. pure water, neutral
            b. coffee, neutral
            c. blood, acid

_____ 8. structural formula, organic compound
            a. carbon; hydrogen
            b. plans, house
            c. isomer, organic molecule

In the space at the left, write the letter of the word or phrase that best completes the statement or answers the question.

_____ 9. The first energy level of an atom holds a maximum of
  a. two electrons.   b. eight electrons.   c. 10 electrons.   d. 18 electrons.

_____ 10. Atoms of a particular element containing different numbers of neutrons are called
  a. elements.   b. ions.   c. crystals.   d. isotopes.

_____ 11. The formula for a molecule of carbon dioxide, CO_2, indicates a total of
  a. two atoms.   b. three atoms.   c. four atoms.   d. six atoms.

Copyright © by the Glencoe Division of Macmillan/McGraw-Hill School Publishing Company   **CHAPTER 3 ASSESSMENT**   **9**

MATTER AND ENERGY

Interpreting and Applying Concepts

Answer the following questions.

1. A Venn diagram is a visual model used to illustrate the similarities and differences that exist among different groups. Compare the properties of biologically important organic compounds listed below by completing the Venn diagram (Figure 3–1). Fill in the nonoverlapping portion of each circle with the letters of the unique characteristics of carbohydrates, lipids, and proteins that are given in the list. Fill in the space where all the circles overlap with the letters of those characteristics that are common to all types of biologically important organic compounds. One of the letters for the overlapping region has been given.

   a. not normally used as a source of energy
   b. contain nitrogen and sulfur
   c. have a ratio of two hydrogen atoms to each oxygen atom
   d. often used as long-term energy reserves
   e. not soluble in water
   f. are organic compounds
   g. number of hydrogen atoms per molecule is much greater than the number of oxygen atoms
   h. often serve as immediate energy sources for life processes
   i. built from amino acids
   j. contain carbon, hydrogen, and oxygen atoms

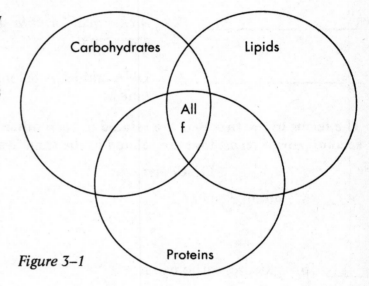

Figure 3–1

2. To which group of biologically important organic compounds do glycogen and cellulose belong?

   _____

3. Compare the structure of glycogen with that of cellulose. _____

   _____

   _____

4. The glucose molecules that make up glycogen are joined by alpha bonds, while cellulose contains beta bonds. Cellulose cannot be digested by enzymes in the human body. If both of these compounds are chains of similar molecules, why can't both be digested by the body?

   _____

   _____

   _____

   _____

Copyright © by the Glencoe Division of Macmillan/McGraw-Hill School Publishing Company

MATTER AND ENERGY

Interpreting and Applying Concepts   continued

Refer to Figure 3–2 to answer the questions that follow.

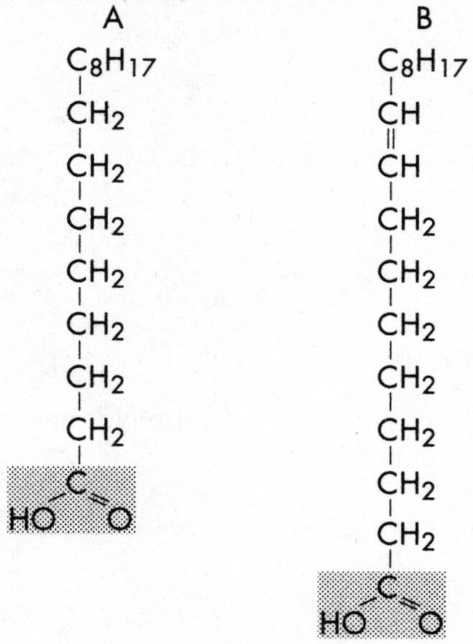

A

C_8H_{17}
CH$_2$
CH$_2$
CH$_2$
CH$_2$
CH$_2$
CH$_2$
CH$_2$
C
HO   O

B

C_8H_{17}
CH
CH
CH$_2$
CH$_2$
CH$_2$
CH$_2$
CH$_2$
CH$_2$
C
HO   O

Figure 3–2

5. What type of biological compounds are illustrated in Figure 3–2? _____

6. How are the structures of the compounds in Figure 3–2 similar? _____

_____

7. How are the structures of the compounds in Figure 3–2 different? _____

_____

8. Classify each compound in Figure 3–2 as either saturated or unsaturated. Provide evidence for your classification. _____

_____

_____

9. Do both compounds illustrated in Figure 3–2 contain the maximum number of hydrogen atoms possible? Explain your answer. _____

_____

_____

10. Certain fats are said to be "polyunsaturated." What would you find upon examining the structural formula of such a compound? _____

_____

Copyright © by the Glencoe Division of Macmillan/McGraw-Hill School Publishing Company

CHAPTER 4 🐛 ASSESSMENT

THE CELL AND ITS ENVIRONMENT

Understanding Concepts

In the space at the left, write TRUE if the statement is true. If the statement is false, change the italicized word or phrase to make it true.

_____ 1. Cell products or wastes are released to the surroundings through the process of *endocytosis.*

_____ 2. The *plasma membrane* of a cell regulates which particles enter and leave the cell.

_____ 3. The diffusion of water into and out of cells across a selectively permeable membrane is called *osmosis.*

_____ 4. *Dynamic equilibrium* is the process of passive transport in which proteins aid the passage of particles across the plasma membrane.

_____ 5. *Phagocytosis* is a form of endocytosis in which liquid droplets are taken in.

Answer the following questions.

6. How do active transport and passive transport differ? _____

_____

_____

7. What is a selectively permeable membrane? _____

_____

8. What is the role of carrier proteins in passive transport? _____

_____

Explain how the terms in each set are related.

9. Diffusion, osmosis, facilitated diffusion

_____

10. Channel proteins, carrier proteins, gate proteins

_____

11. Phagocytosis, pinocytosis, receptor-aided endocytosis

_____

THE CELL AND ITS ENVIRONMENT

Interpreting and Applying Concepts

Refer to Figure 4–1 to answer the questions that follow.

1. What is shown by Part A of the figure?

   _____

   _____

2. What is shown by Part B of the figure?

   _____

   _____

3. Describe how each part of the molecule behaves in

   water. _____

   _____

   _____

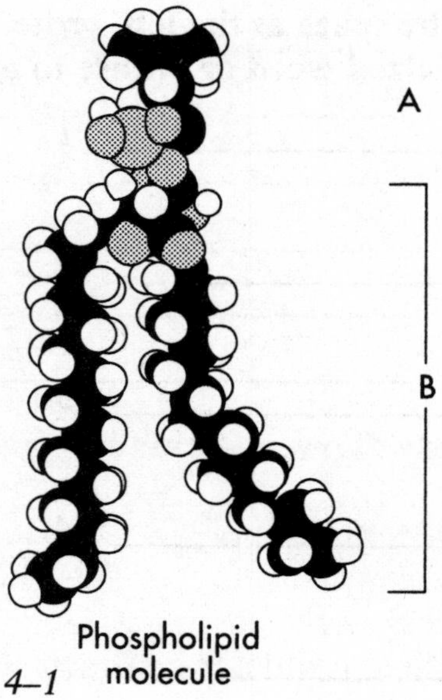

Figure 4–1   Phospholipid molecule

A

B

4. In the space below, sketch the arrangement of molecules, such as the one shown above, in the plasma membrane.

5. Does your sketch accurately depict all substances present in the plasma membrane? If not, identify the missing substances and explain their function.

   _____

   _____

   _____

   _____

    Copyright © by the Glencoe Division of Macmillan/McGraw-Hill School Publishing Company

THE CELL AND ITS ENVIRONMENT

Interpreting and Applying Concepts continued

6. What is there about the structure of the plasma membrane that makes it freely permeable to water? What would happen to the cell if water could not freely pass in and out?

_____

_____

_____

_____

_____

_____

Study Figure 4–2, read the paragraph, and then answer the questions that follow.

A paramecium is a unicellular organism commonly found in ponds and slow-moving streams. These organisms contain structures called contractile vacuoles that are water-storage chambers. Excess water passing through the organism is stored in this structure. When the contractile vacuole is filled, it quickly contracts and pumps the water out of the cell.

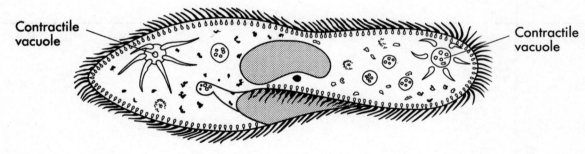

Contractile vacuole   Contractile vacuole

Figure 4–2

7. By what process does water enter a paramecium?

_____

8. What causes water to enter this unicellular organism?

_____

_____

9. Suppose a paramecium were removed from a pond and placed in a container of pure water. What would you observe?

_____

_____

Copyright © by the Glencoe Division of Macmillan/McGraw-Hill School Publishing Company

THE CELL AND ITS ENVIRONMENT

Interpreting and Applying Concepts   continued

10. Do you think it is likely that the paramecium would survive in its new surroundings? Support your answer.

    _____

    _____

    _____

    _____

11. Suppose the paramecium were placed in a container of sea water. What would you observe?

    _____

    _____

    _____

12. How might you adapt the environment of the paramecium described in Question 9 to increase the chances of survival of the organism?

    _____

    _____

   Copyright © by the Glencoe Division of Macmillan/McGraw-Hill School Publishing Company

CHAPTER 5 🐛 ASSESSMENT

INSIDE THE CELL

Understanding Concepts

In the space at the left, write TRUE if the statement is true. If the statement is false, change the italicized word or phrase to make it true.

_____ 1. The sum of all chemical changes in cells is called *metabolism*.

_____ 2. *Lysosomes* are plastids that contain chlorophyll and other pigments needed for photosynthesis.

_____ 3. Cells that do not have a membrane-bound nucleus are called *eukaryotes.*

_____ 4. *Symbiosis* refers to a relationship in which two organisms live closely together.

Explain how the terms in each set are related.

5. chromosomes, nucleus, chromatin

_____

6. microtubules, cytoskeleton, microfilaments

_____

7. bacteria, yeasts, *Amoeba*

_____

8. system, community, organism

_____

In the space at the left, write the letter of the phrase from Column B that best matches the term in Column A.

Column A	Column B
_____ 9. ribosomes	a. control center of a cell
_____ 10. mitochondria	b. delivery system in a eukaryotic cell
_____ 11. nucleus	c. vesicles formed from Golgi bodies
_____ 12. lysosomes	d. protein-making organelles
_____ 13. Golgi bodies	e. powerhouses of a cell

INSIDE THE CELL

Interpreting and Applying Concepts

Answer the following questions.

1. Suppose you were asked to classify an unidentified cell as either prokaryotic or eukaryotic. How would you go about doing this? _____

   _____

   _____

2. Would careful study of an *Amoeba* reveal levels of organization within the organism? Explain your answer. _____

   _____

3. To many scientists, the difference between prokaryotic cells and eukaryotic cells is more important than the difference between plant and animal cells. Do you agree with this

   assessment? Explain your answer. _____

   _____

   _____

Figure 5–1 shows a human blood cell and a human nerve cell. Study the shapes of the cells and answer the questions that follow.

Figure 5–1

　　　　　　　　　　Cell A　　　　　　　　Cell B

4. Identify each cell. Describe the evidence that helped you identify each cell. _____

   _____

   _____

5. Suppose the shapes of the cells illustrated in Figure 5–1 could be reversed. How might the change in shape affect the ability of the cell to do its job?

   _____

   _____

   _____

 Copyright © by the Glencoe Division of Macmillan/McGraw-Hill School Publishing Company

INSIDE THE CELL

Interpreting and Applying Concepts continued

6. Suppose you were asked to classify an unidentified cell as either a plant or animal cell. How would you go about doing this? _____

_____

7. Explain what is meant by the following statement: RNA is to ribosomes as DNA is to

chromosomes. _____

_____

Study Figure 5–2 and answer the questions that follow.

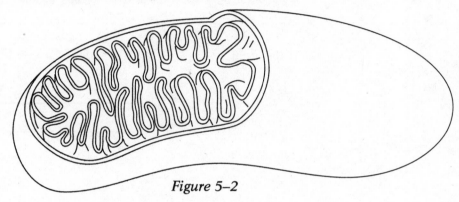

Figure 5–2

8. Identify the kind of cell organelle shown in Figure 5–2. Then explain its function. _____

_____

_____

9. A human liver cell contains many more of the kind of organelle shown in Figure 5–2 than do various other types of cells. What does this indicate about the liver? Why?

_____

_____

_____

10. How is the kind of organelle shown in Figure 5–2 similar to a chloroplast? _____

_____

_____

11. Do bacteria contain the structure shown in Figure 5–2? Explain your answer.

_____

_____

Copyright © by the Glencoe Division of Macmillan/McGraw-Hill School Publishing Company

CHAPTER 6 ☙ ASSESSMENT

THE FLOW OF ENERGY

Understanding Concepts

In the space at the left, write TRUE if the statement is true. If the statement is false, change the italicized word or phrase to make it true.

_____ 1. The *wavelength* of any form of radiant energy is the distance between one wave crest and the next.

_____ 2. *Exergonic* reactions are chemical reactions that require free energy.

_____ 3. Energy for the process of photosynthesis comes from *chemical energy* absorbed by chlorophyll.

_____ 4. In a chloroplast, sugar is synthesized in the *stroma*.

_____ 5. The range of colors that make up white light is known as the *absorption spectrum*.

Explain how the terms in each set are related.

6. lactic acid fermentation, alcoholic fermentation, anaerobic processes

_____

7. chlorophylls, photosynthesis, carotenoids

_____

8. aerobic respiration, glucose, oxygen

_____

9. carbon dioxide, Calvin cycle, sugar, hydrogen ions

_____

10. exergonic reactions, ATP, endergonic reactions

_____

Answer the following question.

11. What are two ways in which photosynthesis and respiration are related?

_____

_____

_____

THE FLOW OF ENERGY

Interpreting and Applying Concepts

Figure 6–1 illustrates the structure of a chloroplast. Use the figure to answer the questions that follow. Write your answers using complete sentences. Use the names of Parts A, B, and C in your answers.

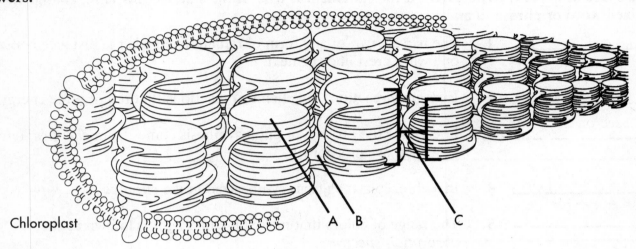

Chloroplast   A B   C

Figure 6–1

1. Explain the relationship that exists between Part A and Part C of the figure.

_____

2. In which part of the figure would you most likely find 3-carbon sugars? Explain your answer.

_____

_____

3. Of the labeled structures shown in the figure, which has a structure similar to that of the cell membrane? Explain how they are alike.

_____

_____

4. Predict the color of Part A of the figure. Provide reasons for your prediction.

_____

_____

_____

5. The structure of a hemoglobin molecule is similar to that of a chlorophyll molecule. The greatest difference between these substances is the presence of an atom of iron in the center of a hemoglobin molecule rather than the magnesium atom found in a chlorophyll molecule. What might this similarity indicate?

_____

Copyright © by the Glencoe Division of Macmillan/McGraw-Hill School Publishing Company

THE FLOW OF ENERGY

Interpreting and Applying Concepts   continued

Refer to Figure 6–2. Answer the questions that follow in complete sentences.

Figure 6–2 **Where Photosynthesis Occurs on Earth**

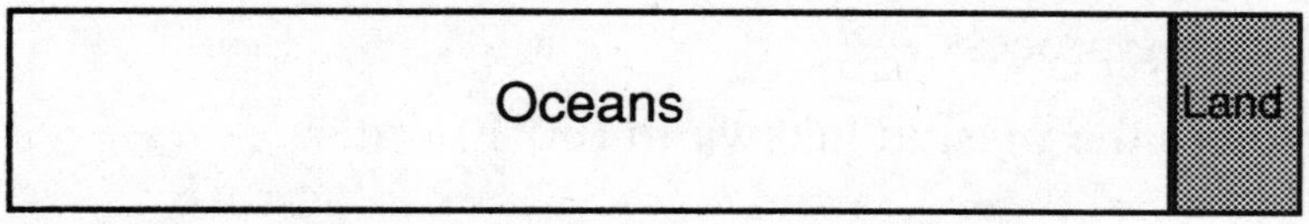

```
┌─────────────────────────────────────────────┬──────┐
│                                              │      │
│                  Oceans                      │ Land │
│                                              │      │
└─────────────────────────────────────────────┴──────┘
```

6. Based on the graph, estimate what percent of all photosynthesis takes place in Earth's oceans.

   _____

7. Explain the danger the world faces if oceans are polluted to the point where populations of photosynthetic organisms are seriously affected.

   _____

   _____

   _____

8. Most cells rely on ATP as a ready energy source to carry out life processes. How do cells get more ATP to replace what has been used?

   _____

   _____

9. What is fermentation? _____

   _____

10. Describe the type of fermentation that could occur in your cells.

   _____

   _____

   _____

11. Compare the products of the type of fermentation described in Question 10 with those of fermentation that occurs in yeast cells.

   _____

   _____

   _____

THE FLOW OF ENERGY

Interpreting and Applying Concepts   continued

12. Yeast cells are living organisms that are dependent on definite temperature ranges. Yeast is dormant at cold temperatures, but begins to activate at about 50°F and is most active between 80°F and 100°F. Yeast cells begin to die at temperatures around 120°F. A bread recipe calls for dissolving yeast in warm (85°F) water. Predict the effect each of the following actions would have on the dough-rising process.

   a. The baker dissolves the yeast in lukewarm (60°F) water.

   _____

   _____

   b. The baker dissolves the yeast in boiling water.

   _____

   _____

13. In an effort to reduce energy needs, the alternative fuel gasohol was developed. This mixture consists of 90 percent gasoline and 10 percent ethanol. Explain how yeast could be used to make gasohol.

   _____

   _____

14. In a particular experiment, scientists used water molecules that contained the radioactive isotope oxygen-18. What do you suppose would occur if this water were added to the soil of a potted green plant?

   _____

   _____

Copyright © by the Glencoe Division of Macmillan/McGraw-Hill School Publishing Company

Name_____ Date_____ Class_____

CHAPTER 7 🐌 ASSESSMENT

CELLULAR REPRODUCTION

Understanding Concepts

In the space at the left, write the letter of the word or phrase that best completes the statement or answers the question.

_____ 1. In your body, cells that could *not* be undergoing cell division are
  a. nerve cells.   b. skin cells.   c. red blood cells.   d. liver cells.

_____ 2. Prokaryotes produce daughter cells by the process of
  a. meiosis I.   b. meiosis II.   c. binary fission.   d. mitosis.

_____ 3. In organisms that reproduce by sexual reproduction, the 2*n* number of chromosomes is restored by
  a. cell division.   b. binary fission.   c. spore formation.   d. fertilization.

_____ 4. The two processes that provide for variation and genetic recombination among offspring are
  a. fertilization and mitosis.   c. cell division and mitosis.
  b. fertilization and meiosis.   d. cell division and meiosis.

_____ 5. Pasteur's work led directly to the
  a. cell theory.   c. theory of biogenesis.
  b. theory of continuance.   d. theory of spontaneous generation.

_____ 6. You know you are observing plant cells undergoing mitosis if you observe
  a. cell plates.   c. tetrads.
  b. centrioles.   d. endoplasmic reticulum.

_____ 7. Meiosis in humans results in
  a. one sperm cell.   c. four identical sperm cells.
  b. one egg cell.   d. four identical egg cells.

_____ 8. In plants and some algae and fungi, meiosis results in
  a. haploid spores.   b. gametes.   c. microtubules.   d. flagella.

In the space at the left, write TRUE if the statement is true. If the statement is false, change the italicized word or phrase to make it true.

_____ 9. The union of a sperm cell and an egg cell results in a *gamete*.

_____ 10. The region at which the two strands of a chromosome are joined is the *centromere*.

_____ 11. The process that ensures that each daughter cell has the same number and kind of chromosomes as the parent cell is *meiosis*.

Copyright © by the Glencoe Division of Macmillan/McGraw-Hill School Publishing Company

CELLULAR REPRODUCTION

Interpreting and Applying Concepts

Use Figure 7–1 showing the phases of the cell cycle to answer the questions.

1. What part of the cycle does *X* represent?

   _____

2. What is the most important process
   to occur during that part of the cycle?

   _____

3. About what percentage of the cell cycle is
   taken up by mitosis?

   _____

4. Suppose a drug that prevents proper
   development of cell structures in the last few
   hours of interphase is used. What would

   be the result? _____

   _____

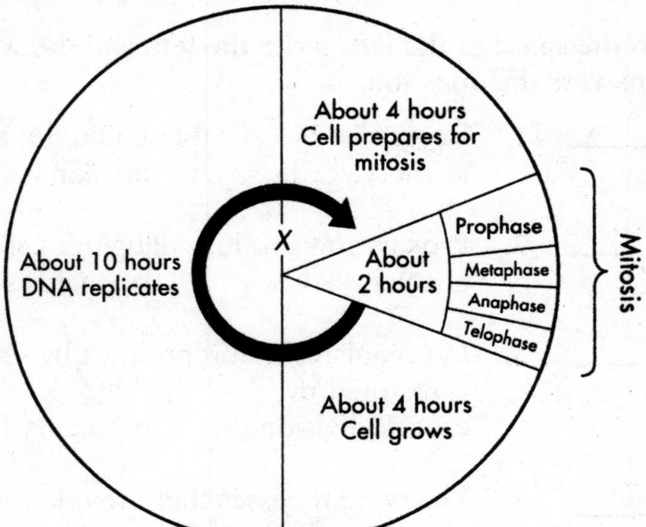

Figure 7–1   A 20-Hour Cell Cycle

When meiosis I occurs following fertilization in an organism with four chromosomes, two alternate arrangements of the chromosomes are possible. Refer to Figure 7–2 and answer the question that follows.

Figure 7–2

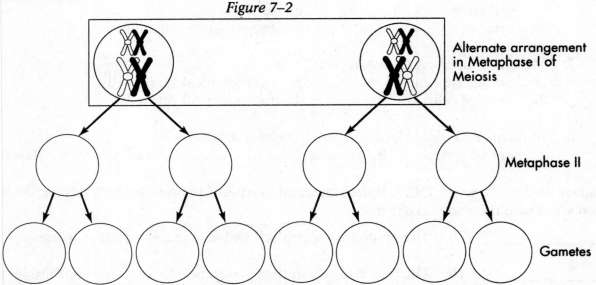

5. Complete the figure. Draw the chromosomes in each cell in metaphase II to show the possible arrangements of the chromosomes. Then draw the chromosomes in the gametes to show the possible combinations of chromosomes. (You may use colored pencils.)

Copyright © by the Glencoe Division of Macmillan/McGraw-Hill School Publishing Company

CELLULAR REPRODUCTION

Interpreting and Applying Concepts   continued

The diagrams in Figure 7–3 show cells from the same eukaryotic animal cell in various stages of mitosis. Use the diagrams to answer the questions.

1.

2.
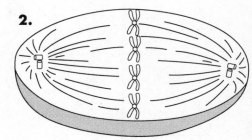

3.
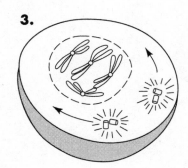

4.
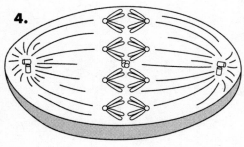

5.
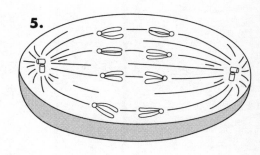

Figure 7–3

6. What is the chromosome number for the cells in the diagram? _____

7. What is the structure labeled *X* in diagram 2? _____

8. Which diagram correctly represents a cell in anaphase? _____

9. How many chromatids are shown in diagram 2? _____

10. Which diagram shows a stage of mitosis incorrectly? _____

11. Which structure indicates that you are observing mitosis in

    an animal cell? _____

12. List the proper order for the four *correct* diagrams. _____

Copyright © by the Glencoe Division of Macmillan/McGraw-Hill School Publishing Company

Name _____   Date _____   Class _____

HEREDITY

Understanding Concepts

Write the letter of the word or phrase that best completes the statement or answers the question.

1. When a pure tall pea plant is crossed with a pure short pea plant,

   the _____ trait appears in the F_1 generation.

2. The phenotypic ratio obtained when two organisms that are heterozygous for two different

   traits are crossed is _____ .

3. The genotypic ratio obtained when an *Rr* organism is crossed with an *rr* organism is

   _____ . (Include letters.)

4. When two pea plants that are heterozygous for green pods are crossed, the genotypic ratio that

   results is _____. (Include letters.)

In the space at the left, write the term in parentheses that correctly completes the statement.

_____ 5. According to the laws of probability, the larger the number of trials, the (*closer to, farther from*) the expected ratio the results will be.

_____ 6. Human skin color, which shows continuous variation, appears to be inherited through (*multiple genes, modifier genes*).

_____ 7. An organism with two unlike alleles for the same trait is (*homozygous, heterozygous*).

_____ 8. The environment in which an organism develops (*can, cannot*) affect the expression of its genetic traits.

In the space at the left, write the letter of the phrase from Column B that best matches the term in Column A.

Column A	Column B
_____ 9. autosomes	**a.** An individual with two similar alleles for the same trait
_____ 10. codominance	**b.** The science of heredity
_____ 11. genetics	**c.** Genes for different traits segregate without regard for each other
_____ 12. homozygous	**d.** The physical appearance of an organism
_____ 13. incomplete dominance	**e.** Two alleles result in an intermediate third phenotype
_____ 14. phenotype	**f.** Two alleles expressed equally in the offspring
_____ 15. principle of independent assortment	**g.** Chromosomes not involved in determining the sex of an individual

Copyright © by the Glencoe Division of Macmillan/McGraw-Hill School Publishing Company

HEREDITY

Interpreting and Applying Concepts

When a white, flat squash (*WWFF*) is crossed with a yellow, round squash (*wwff*), all the offspring are white and flat (*WwFf*). Suppose two of these offspring are crossed. The partial results in the F₂ generation are shown in the Punnett square in Figure 8-1. Study Figure 8–1 and then answer the questions that follow.

Figure 8–1

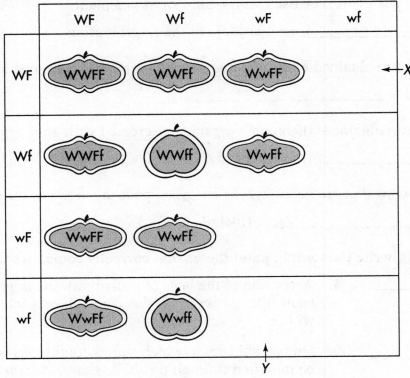

_____ 1. How many white, round squash would be indicated by the complete Punnett square?

_____ 2. Draw the squash that would appear in the square labeled *x*. Be sure to follow the pattern for shape and color indicated.

_____ 3. What is the genotype of the squash in the square labeled *x*?

_____ 4. Draw the squash that would appear in the square labeled *y*.

_____ 5. What is the phenotype of the squash in the square labeled *y*?

_____ 6. Give the ratio for all the phenotypes that would be indicated by the complete Punnett square. (Include traits in answer.)

   Copyright © by the Glencoe Division of Macmillan/McGraw-Hill School Publishing Company

HEREDITY

Interpreting and Applying Concepts   continued

Assume that a mother with blood type *A* has a child of blood type *O*. Answer the following questions.

7. What is the genotype of the mother? How do you know? _____

_____

8. What blood types could the father *not* be? _____

9. If the father was type *A* would he be I^AI^A or I^Ai? _____

10. If the father was type *B*, would he be I^BI^B or I^Bi? _____

11. Must the father be type *O*? Why or why not? _____

_____

Answer the following questions.

12. Draw and label a Punnett square containing the human sex chromosomes to show the inheritance of maleness and femaleness in humans. (Use the appropriate letters for the chromosomes.)

13. Use the data from the Punnett square to explain why approximately half of all infants born are male and half are female. _____

_____

_____

_____

14. Use the data from the Punnett square to help explain why sex-linked traits, such as color blindness, may appear occasionally in men but very rarely in women.

_____

_____

_____

Copyright © by the Glencoe Division of Macmillan/McGraw-Hill School Publishing Company

Name _____ Date _____ Class _____

CHEMISTRY OF THE GENE

Understanding Concepts

In the space at the left, write the letter of the word or phrase that best completes the statement or answers the question.

_____ 1. Each of these scientists was involved in identifying the transforming principle *except*
    a. Griffith.   **b.** Avery.   **c.** Hershey and Chase.   **d.** Watson and Crick.

_____ 2. The genetic code in DNA depends on the order of the
    a. nucleotides.   **b.** pyrimidines.   **c.** major grooves.   **d.** phosphate groups.

_____ 3. Protein synthesis involves the use of
    a. 6 nucleotides.   **c.** 20 polypeptides.
    b. 16 amino acids.   **d.** 64 codons.

_____ 4. A nucleotide consists of each of the following *except* a(n)
    a. deoxyribose sugar.   **c.** amino acid.
    b. phosphate group.   **d.** nitrogen-containing base.

_____ 5. In the ladder-like model of DNA, the uprights consist of
    a. nucleotides.   **c.** sugar and phosphate groups.
    b. nitrogen bases.   **d.** ATP.

_____ 6. In a eukaryote, replication of DNA *always* begins
    a. at one end.   **c.** at several places along the DNA.
    b. between two codons.   **d.** within one of the nucleotides.

_____ 7. During replication of DNA, a single mistake in a nucleotide
    a. is always fatal.   **c.** occurs in 1 of 1000 nucleotides.
    b. is usually self-repairing.   **d.** prevents variation in genes.

_____ 8. One segment of nucleotides along a DNA molecule forms a
    a. purine.   **b.** pyrimidine.   **c.** phage.   **d.** gene.

In the space at the left, write the term in parentheses that best completes the statement.

_____ 9. In DNA, each set of three nitrogen bases representing an amino acid is a (*nucleotide, codon*).

_____ 10. After replication, each new strand of DNA, compared with one of the two parent strands, is (*a complement, identical*).

_____ 11. In cells, the long chains formed of hundreds of amino acids are (*polypeptides, phosphates*).

_____ 12. If one strand of DNA has the nitrogen bases TATGC, the other strand has the bases (*TATGC, ATACG*).

Copyright © by the Glencoe Division of Macmillan/McGraw-Hill School Publishing Company

CHEMISTRY OF THE GENE

Interpreting and Applying Concepts

Read the paragraph and answer the questions that follow.

In 1958, the replication model was verified by the experiments of Meselson and Stahl. They grew *E. coli* bacteria in a medium containing N^{15} (heavy form of nitrogen). After many generations, all their DNA molecules contained only N^{15}. Then these bacteria were grown on a medium containing N^{14} (ordinary light nitrogen). After one cell division, the DNA in the bacteria was halfway in weight between N^{15} and N^{14}. Each new DNA molecule had one strand containing N^{15} and one strand containing N^{14}. Finally, these bacteria divided again on the N^{14} medium.

1. Complete Figure 9-1 to show the types of DNA in the four cells that developed in the second generation.

Figure 9–1

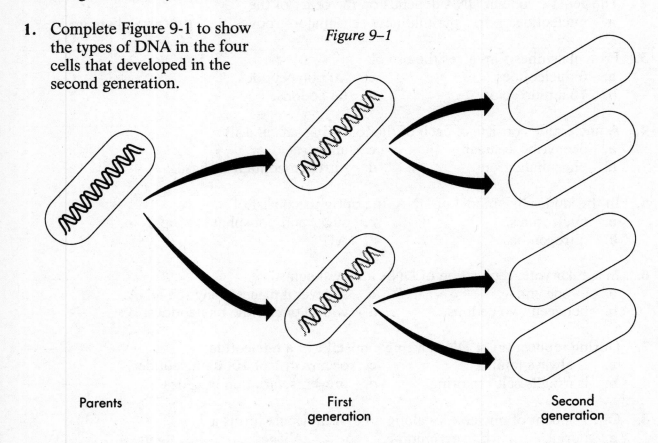

Parents First generation Second generation

2. Explain your answer to Exercise 1.

_____

_____

_____

_____

_____

 Copyright © by the Glencoe Division of Macmillan/McGraw-Hill School Publishing Company

CHEMISTRY OF THE GENE

Interpreting and Applying Concepts continued

Study the codons of DNA in Table 9–1 to answer the questions.

Table 9–1

First Base in Codon	Second Base in Codon				Third Base in Codon
	A	**G**	**T**	**C**	
A	phenylalanine	serine	tyrosine	cysteine	**A**
	phenylalanine	serine	tyrosine	cysteine	**G**
	leucine	serine	stop	stop	**T**
	leucine	serine	stop	tryptophan	**C**
G	leucine	proline	histidine	arginine	**A**
	leucine	proline	histidine	arginine	**G**
	leucine	proline	glutamine	arginine	**T**
	leucine	proline	glutamine	arginine	**C**
T	isoleucine	threonine	asparagine	serine	**A**
	isoleucine	threonine	asparagine	serine	**G**
	isoleucine	threonine	lysine	arginine	**T**
	methionine	threonine	lysine	arginine	**C**
C	valine	alanine	aspartate	glycine	**A**
	valine	alanine	aspartate	glycine	**G**
	valine	alanine	glutamate	glycine	**T**
	valine	alanine	glutamate	glycine	**C**

3. What does each codon represent? _____

4. What are the three letters that make up each stop codon? _____

5. What two amino acids are represented by only one codon each? _____

_____

6. In codons for the same amino acids, the first two nitrogen bases are usually the same. Which
    amino acids form an exception to this rule? _____

7. Codon TAC has a dual function. One function is to signal the start of protein synthesis. What
    is the other function? _____

CHEMISTRY OF THE GENE

Interpreting and Applying Concepts   continued

Read the paragraph and answer the questions that follow.

In building their model of DNA, Watson and Crick built on the work of other scientists who had studied various aspects of the structure of DNA. However, Watson and Crick had to answer several major questions about the structure of DNA in order to develop their model. This now-famous model accounted for the apparent uniform width of DNA and explained a great mystery: How could DNA copy itself?

8. Linus Pauling proposed a three-stranded model of DNA. To Watson and Crick, the width of the helix suggested that DNA was two-stranded. Why else did they reason it was two stranded?

_____

_____

9. Why did Watson and Crick call their model a double helix?

_____

_____

_____

10. At first, Watson and Crick believed that bases paired with like bases — A with A, C with C, and so forth. Why was this idea disproved?

_____

_____

11. Originally, Watson and Crick placed the sugar-phosphate chain on the inside of the molecule. Why did they move it to the outside?

_____

_____

_____

12. In 1953, Watson and Crick surprised the scientific community when they published a report of their model of DNA in the journal *Nature.* Why has this model become a cornerstone of modern molecular biology?

_____

_____

Copyright © by the Glencoe Division of Macmillan/McGraw-Hill School Publishing Company

CHAPTER 10 🐛 ASSESSMENT
FROM GENES TO PROTEINS

Understanding Concepts

In the space at the left, write the term in parentheses that best completes the statement.

_____ 1. RNA, unlike DNA, contains the nitrogen base (*uracil, thymine*).

_____ 2. The process of transferring the genetic information from DNA to RNA is called (*translation, transcription*).

_____ 3. A gene mutation that affects only one amino acid is a (*point, chromosome*) mutation.

_____ 4. Nucleotide sequences that do not code for amino acids are (*exons, introns*).

_____ 5. During protein synthesis, the (*codons, ribosomes*) move along the mRNA strand.

_____ 6. In bacteria, DNA that has been altered by insertion of a foreign gene is called (*recombinant, template*) DNA.

_____ 7. Scientists estimate that approximately (*one percent, 90 percent*) of DNA in eukaryotic cells codes for polypeptides.

_____ 8. Mutations may affect an entire population of organisms if they occur in (*body, sex*) cells.

_____ 9. Barbara McClintock received a Nobel Prize about 40 years after her discovery of (*pseudogenes, jumping genes*).

_____ 10. Abnormal cells that can migrate from one organ to another are (*gene-cloned, malignant*) cells.

_____ 11. Plants that contain genetic materials from two species are (*transformed, transgenic*) plants.

_____ 12. RNA is usually composed of a (*single, double*) chain of nucleotides.

_____ 13. In protein synthesis, amino acids are brought to the ribosome by (*transfer, messenger*) RNA.

_____ 14. The sugar present in RNA is (*ribose, deoxyribose*).

In the space at the left, write the letter of the phrase from Column B that best matches the term in Column A.

Column A	Column B
_____ 15. anticodon	**a.** three bases at one end of a tRNA molecule
_____ 16. clones	**b.** changes in the genes
_____ 17. mutations	**c.** genes that mutate and cause cancer
_____ 18. oncogenes	**d.** small circles of DNA in bacteria
_____ 19. plasmids	**e.** genetically identical copies of organisms

FROM GENES TO PROTEINS

Interpreting and Applying Concepts

Study Figure 10–1, in which four stages of protein synthesis, labeled 1–4, are shown out of order, and then answer the following questions.

Figure 10–1

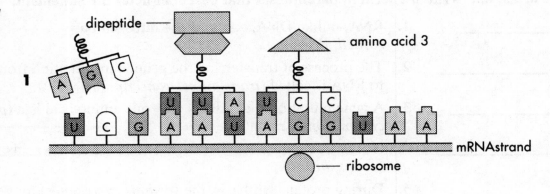

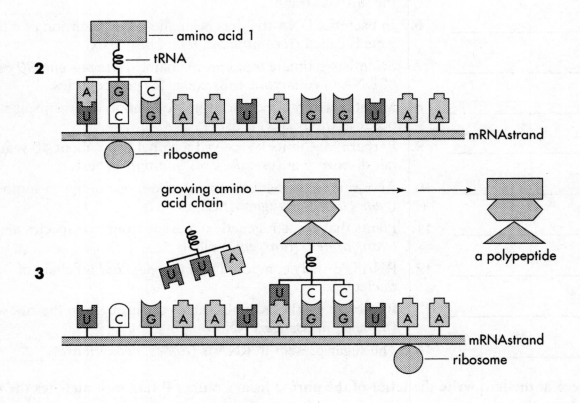

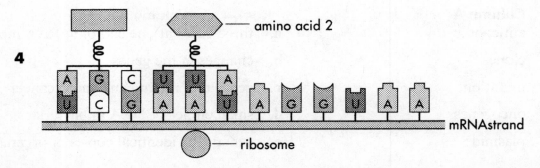

  Copyright © by the Glencoe Division of Macmillan/McGraw-Hill School Publishing Company

FROM GENES TO PROTEINS

Interpreting and Applying Concepts   continued

1. Use the numbers of the stages to list them in proper order. _____

2. How many different tRNA molecules are shown in the figure? Explain your answer.

_____

3. How many different mRNA strands are shown in the figure? Why do you say so?

_____

4. Where is a stop codon located? How do you know? _____

_____

5. What process had to precede the portion of protein synthesis shown in the figure? Explain why.

_____

_____

Study Figure 10–2, which shows four kinds of chromosome mutations, and then answer the following questions.

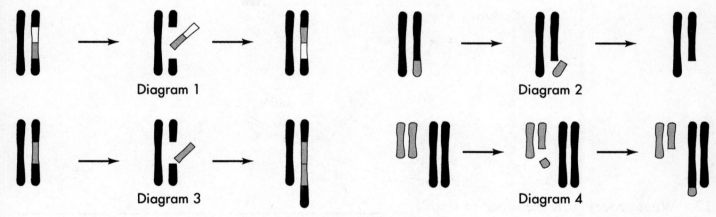

Diagram 1   Diagram 2

Diagram 3   Diagram 4

Figure 10–2

6. What type of mutation is shown in diagram 1? _____

7. What type of mutation is shown in diagram 2? _____

8. What type of mutation is shown in diagram 3? _____

9. Diagram 4 shows translocation, another type of chromosome mutation. Describe what

happens during translocation. _____

10. What types of chromosomal mutations involve change in one chromosome of a pair?

_____

Copyright © by the Glencoe Division of Macmillan/McGraw-Hill School Publishing Company

FROM GENES TO PROTEINS

Interpreting and Applying Concepts   continued

11. Are the chromosome pairs involved in translocation homologous or nonhomologous?

_____

Study Figure 10–3, which shows RNA processing, and then answer the following questions.

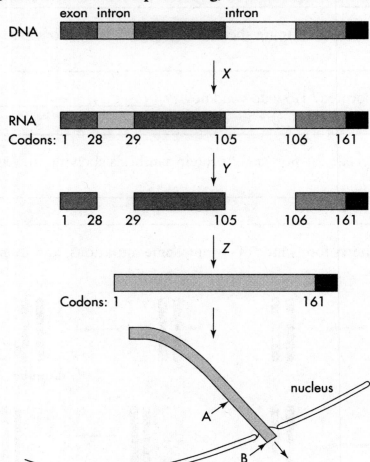

Figure 10–3

12. What process is represented at step X? _____

13. How many amino acids long is this polypeptide? _____

14. What is happening at step Y? _____

15. How many exons, or coding segments, are present? _____

16. What is happening at step Z? _____

17. What kind of RNA is shown at A? _____

18. Describe what is happening at B. _____

_____

Copyright © by the Glencoe Division of Macmillan/McGraw-Hill School Publishing Company

CHAPTER 11 ❧ ASSESSMENT

HUMAN GENETICS

Understanding Concepts

In the space at the left, write the term in parentheses that best completes the statement.

_____ 1. The process by which parents are given information about chances of their unborn child having an inherited disorder is (*genetic counseling, gene therapy*).

_____ 2. The presence of three of one kind of chromosome in a cell is called (*trisomy, nondisjunction*).

_____ 3. Most human genetic disorders result from a (*homozygous recessive, heterozygous dominant*) genotype.

_____ 4. Cystic fibrosis, Huntington's disease, and sickle-cell anemia are the result of (*missing, mutant*) alleles.

In the space at the left, write the letter of the word or phrase that best completes the statement or answers the question.

_____ 5. A fetus is viewed directly through an endoscope in
    a. amniocentesis.   **c.** echograms.
    b. fetoscopy.   **d.** chorionic villus biopsy.

_____ 6. In parts of Africa where malaria is present, blacks who are heterozygous for sickle-cell hemoglobin tend to live longer than those with other genotypes. These individuals
    a. suffer from the disease.   **c.** often die of malaria.
    b. have heterozygous superiority.   **d.** have severe pain.

_____ 7. Huntington's disease is caused by
    a. an extra chromosome.   **c.** a single dominant gene.
    b. two recessive genes.   **d.** a missing chromosome.

_____ 8. An autosomal recessive disorder in which degeneration of the nervous system begins before the end of a baby's first year is
    a. Tay-Sachs disease.   **c.** diabetes mellitus.
    b. Down syndrome.   **d.** leukemia.

_____ 9. Each of the following is related to cystic fibrosis *except*
    a. abnormal cell membrane protein.
    b. most common inherited disorder among whites.
    c. deteriorating brain cells.
    d. accumulation of mucus in lungs.

_____ 10. If one parent has a disorder caused by a dominant gene, the chance of the offspring having the disorder
    a. is 0 percent.   **c.** is 100 percent.
    b. is 50 percent.   **d.** cannot be determined.

Copyright © by the Glencoe Division of Macmillan/McGraw-Hill School Publishing Company

HUMAN GENETICS

Interpreting and Applying Concepts

One type of antigen in all human blood is the Rh factor. It is present in about 85 percent of the people in the United States, who are classified as Rh-positive (Rh$^+$). People whose blood does not contain the Rh factor are classified as Rh negative (Rh$^-$). Study Table 11–1, which indicates the inheritance of the Rh factor within 98 families. Then answer the questions that follow.

Table 11–1

Parent	Number of families	Children: Rh$^+$	Rh$^-$	Percentage Rh-negative
Rh$^+$ × Rh$^+$	69	217	15	6.5
Rh$^+$ × Rh$^-$	21	55	24	30.4
Rh$^-$ × Rh$^-$	8	0	37	100.0

1. What percentage of children born to two Rh-positive parents were Rh-negative? _____

2. How many Rh-positive children were born to families in which one parent was Rh-positive and the other Rh-negative? _____

3. What percentage of children born to families in which both parents were Rh–negative were Rh-positive? _____

4. Is the presence of the Rh factor a dominant or a recessive trait? Use the table to explain.

   _____

   _____

   _____

   _____

5. In families having one Rh-positive parent and one Rh-negative parent, was the Rh-positive parent homozygous or heterozygous? How do you know? _____

   _____

   _____

Copyright © by the Glencoe Division of Macmillan/McGraw-Hill School Publishing Company

HUMAN GENETICS

Interpreting and Applying Concepts   continued

Read the paragraph and then answer the following questions.

Some individuals contain a gene that causes blue sclera, a condition in which the whites of the eyes appear bluish. Anyone who possesses the gene, whether homozygous or heterozygous, is expected to show blue sclera. But it has been found that only about nine out of ten people with this gene show blue sclera. Among those showing the phenotype, the blue color ranges from pale whitish blue to dark blackish blue.

Scientists study this condition for penetrance and expressivity. Penetrance is the percentage of individuals carrying the gene that actually show it. Expressivity is the way the phenotype is expressed.

6. Is the gene for blue sclera dominant or recessive? How do you know?

    _____

    _____

7. If two individuals homozygous for blue sclera marry, what percentage of their offspring would

    you expect to have the gene? _____

8. What percentage would you expect to actually show blue sclera? Why do you say so?

    _____

9. What is the expressivity of the phenotype for blue sclera?

    _____

    _____

Study Figure 11–1. Then read the paragraph and answer the questions that follow.

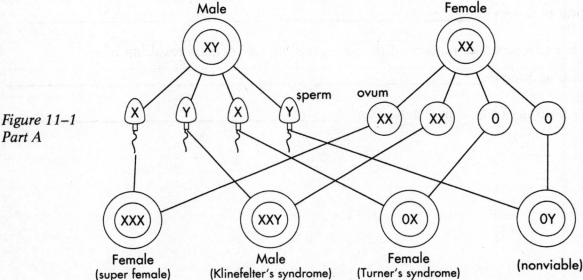

*Figure 11–1
Part A*

HUMAN GENETICS

Interpreting and Applying Concepts   continued

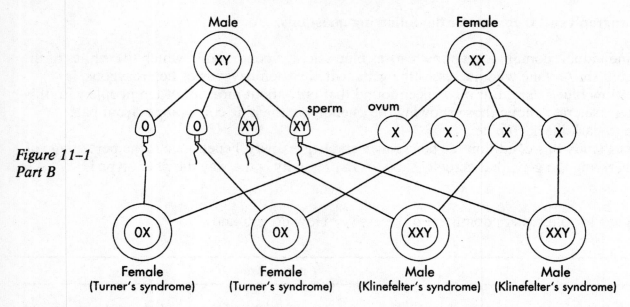

Figure 11–1
Part B

Figure 11–1 shows the result of nondisjunction, a failure of paired chromosomes to separate, in sex chromosomes of male and females. Part A shows the result of nondisjunction in females; Part B shows the result in males.

An individual with Turner's syndrome is a female, but does not develop normally and is sterile. She may be mildly retarded. An individual with Klinefelter's syndrome may appear to be a normal male, but is probably sterile and may show some mental retardation. Although a super female has an extra *X* chromosome, she does not have super ability and may be sterile.

10. What is the sex chromosome pattern of an individual with Klinefelter's syndrome? _____

11. What is the sex chromosome pattern of an individual with Turner's syndrome? _____

12. What is the sex chromosome pattern of the four possible egg cells resulting from

    nondisjunction in females? _____

13. What is the sex chromosome pattern of the four possible sperm cells resulting from

    nondisjunction in males? _____

14. What is the meaning of the O in the diagrams? _____

Copyright © by the Glencoe Division of Macmillan/McGraw-Hill School Publishing Company

CHAPTER 12 ✒ ASSESSMENT
EVOLUTION

Understanding Concepts

In the space at the left, write the letter of the word or phrase that best completes the statement or answers the question.

_____ 1. The forelegs of a cat and bat are examples of
     a. comparative biochemistry.   **c.** analogous structures.
     b. homologous structures.   **d.** comparative embryology.

_____ 2. The similarity among the blood proteins of all the mammals may be taken as evidence for evolutionary relationships based upon
     a. comparative anatomy.   **c.** geographic distribution.
     b. comparative embryology.   **d.** comparative biochemistry.

_____ 3. Which conclusion may be drawn when comparing fossils found in previously undisturbed strata of sedimentary rock?
     a. Fossils in the upper strata are younger than those in the lower strata.
     b. Fossils in the upper strata are older than those in the lower strata.
     c. Fossils in the upper strata are generally less complex than those in the lower strata.
     d. There are no fossils in the upper strata that resemble those in the lower strata.

_____ 4. Which of the following was *not* part of Darwin's proposals?
     a. There is overpopulation in nature.
     b. There is variation in the individuals of the population.
     c. The variations are caused by mutations.
     d. There is natural selection.

Answer the following questions.

5. Why is the fossil evidence incomplete? _____

_____

_____

6. What types of organisms have humans selectively bred? _____

_____

7. Do bacteria exposed to penicillin mutate in response to the penicillin or do they already

possess the mutation? _____

_____

_____

8. What four elements necessary for life were contained in the gases that made up Earth's early

atmosphere? _____

Copyright © by the Glencoe Division of Macmillan/McGraw-Hill School Publishing Company

EVOLUTION

Interpreting and Applying Concepts

Answer the following questions in complete sentences.

1. Darwin calculated that, if a single breeding pair of elephants reproduced and all their offspring survived and reproduced, in 750 years there would be a standing population of 19 million elephants. In Africa and Asia, where elephants live in the wild, the total number of elephants is currently less than one million. Explain this in terms of natural selection.

_____

_____

_____

_____

_____

_____

_____

2. Thomas Malthus, in 1798, warned that the human population was increasing so rapidly that it would soon be impossible to feed all of Earth's inhabitants. Darwin noted that food supply and other factors would hold populations in check. Human populations continue to grow at rapid rates in many countries of the world. Why do you think humans have avoided the pressures of selection indicated by Malthus and Darwin?

_____

_____

_____

_____

_____

3. How does selective breeding influence the process of evolution?

_____

_____

_____

_____

_____

EVOLUTION

Interpreting and Applying Concepts   continued

Table 12–1 gives a partial amino acid sequence in hemoglobin proteins of humans, gorillas, and horses. Answer the following questions.

Table 12–1

HUMAN	Gly	Lys	Val	Asp	Val	Asp	Glu	Val	Gly	Gly	Glu
GORILLA	Gly	Lys	Val	Asp	Val	Asp	Glu	Val	Gly	Gly	Glu
HORSE	Asp	Lys	Val	Asp	Glu	Glu	Glu	Val	Gly	Gly	Glu

HUMAN	Lys	Leu	His	Val	Asp	Pro	Glu	Asp	Phe	Arg	Leu
GORILLA	Lys	Leu	His	Val	Asp	Pro	Glu	Asp	Phe	Leu	Leu
HORSE	Lys	Leu	His	Val	Asp	Pro	Glu	Asp	Phe	Arg	Leu

4. Circle those instances in Figure 12–1 when the amino acids are not the same among the three organisms.

5. In this shortened sequence, how many times were there differences between the sequence of human and gorilla amino acids? _____

6. How many times were there differences between the human and the horse?

_____

7. From this information and your knowledge of biology, which two organisms do you think evolved from the most recent common ancestor? _____

8. Give reasons for supporting or rejecting the following statement: Evolutionary relationships are stronger between organisms that have close biochemical similarities than between organisms that do not have close biochemical similarities.

_____

_____

_____

_____

_____

Copyright © by the Glencoe Division of Macmillan/McGraw-Hill School Publishing Company

EVOLUTION

Interpreting and Applying Concepts   continued

Look at Figure 12–1 and answer the question.

| **Elephant** | **Wooly Mammoth** |

Figure 12–1

9. According to biochemical tests, there is a close evolutionary link between the wooly mammoth, which roamed North America, Asia, and Europe thousands of years ago, and the modern elephant (Figure 12–1). What changes do you think took place for the elephant to have evolved from the ancestral mammoth?

_____

_____

_____

_____

_____

_____

_____

_____

   Copyright © by the Glencoe Division of Macmillan/McGraw-Hill School Publishing Company

CHAPTER 13 ❧ ASSESSMENT

ADAPTATION AND SPECIATION

Understanding Concepts

In the space at the left, write TRUE if the statement is true or FALSE if the statement is false.

_____ 1. The leaves on a tree are adapted for carrying on photosynthesis.

_____ 2. Any variation found in a plant or animal is considered an adaptation for survival of the individual.

_____ 3. Adaptations are the result of chance mutations that accumulate over a long period of time.

_____ 4. Horses and donkeys are considered to be in the same species.

_____ 5. A tetraploid (4n) plant is an example of a polyploid organism.

_____ 6. *Homo sapiens* are believed to have first appeared on Earth about 1.8 million years ago.

_____ 7. The idea that evolution occurs through slow and steady adaptive changes in populations is known as convergent evolution.

_____ 8. In the long process of evolution of the eye, a probable early step was the development of a group of light sensitive cells in simple organisms.

_____ 9. The poisonous venom of a snake is a physiological adaptation.

_____ 10. The oldest fossils of human ancestors were discovered in Europe.

Answer the following questions.

11. What is speciation? _____

12. How does geographic isolation influence speciation? _____

_____

_____

13. What are three adaptations common to all primates? _____

_____

14. Describe a behavioral adaptation of birds. _____

_____

15. Compare gradualism and punctuated equilibrium. _____

_____

_____

ADAPTATION AND SPECIATION

Interpreting and Applying Concepts

Answer the following questions in complete sentences.

1. Describe the evolution of finches on the Galapagos Islands, using the following terms in your answer: reproductive isolation, ancestral population, geographic isolation, adaptive radiation, gene pools.

_____

_____

_____

_____

_____

_____

_____

_____

_____

2. Describe how a fire in a field could cause genetic drift in the populations of weeds that were left to reproduce.

_____

_____

_____

_____

_____

_____

3. If you were given a large box of primate skeletons, what are some factors that would help you classify them according to how each fit into a sequence of evolution of primates?

_____

_____

_____

_____

Copyright © 1994 by the Glencoe Division of Macmillan/McGraw-Hill Publishing Company

ADAPTATION AND SPECIATION

Interpreting and Applying Concepts   continued

In the presumed evolution of the horse, paleontologists have been able to construct a reasonable picture of the probable sequence of ancestors to the modern horse. Study the time line in Figure 13–1 and the descriptions of early ancestors below it. In the space at the left of each description, write the letter that indicates about how many years ago each ancestor lived.

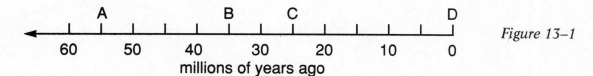

Figure 13–1

_____ **4.** This ancestor was much larger than a dog and both front and hind feet had three toes.

_____ **5.** The modern horse is large and has one toe on each foot.

_____ **6.** This ancestor was a grazing animal with three toes on each foot, the center toe being larger than the other two. The two smaller toes did not reach the ground.

_____ **7.** This ancestor was about the size of a fox terrier. It had four toes on each front foot and three toes on each hind foot. It fed on trees and bushes.

Answer the following questions in complete sentences.

8. Describe what selective pressures favored the evolution of the horse to its present form.

_____

_____

_____

_____

9. Figure 13–2 is a picture of the extinct dodo bird. Compare the dodo to any kind of bird that you know that can fly. Describe the adaptations found in the modern bird that were not evident in the dodo.

Dodo (extinct)

_____

_____

_____

_____

_____

_____

Figure 13–2

Copyright © 1994 by the Glencoe Division of Macmillan/McGraw-Hill Publishing Company

CHAPTER 14 🐾 ASSESSMENT

CLASSIFICATION

Understanding Concepts

In the space at the left, write the letter of the word or phrase that best completes the statement or answers the question.

_____ 1. The scientific name *Homo sapiens* indicates that organisms given this name
    a. belong to the species *Homo.*   **c.** belong to the class *Homo.*
    b. belong to the genus *Homo.*   **d.** belong to the phylum *Homo.*

_____ 2. Which kind of evidence would help most to classify an organism?
    a. the organism's habitat
    b. the evolutionary relationships of the organism
    c. the organism's eating habits
    d. the function of the organism's body parts

_____ 3. Which is true of organisms classified in the same genus?
    a. They are in the same order but in different families.
    b. They belong to the same species but could be in different phyla.
    c. They must be in the same kingdom but could be in different phyla.
    d. They are in the same kingdom but may be of different species.

_____ 4. Which of the following is true?
    a. The giant panda and the red panda are both related to raccoons.
    b. The giant panda and the red panda are both related to bears.
    c. The giant panda is related to bears and the red panda is related to raccoons
    d. The giant panda is related to raccoons and the red panda is related to bears.

Write the word or phrase that best completes the statement.

5. The science of classification is called _____ .

6. The most inclusive taxon is the _____.

7. In their classification, the cat and the dog diverge at the level of the _____.

8. Carolus Linnaeus introduced a system of naming organisms called_____ .

9. Modern classification uses techniques of molecular biology, in particular the analysis of

proteins and _____ .

In the space at the left, write TRUE if the statement is true or FALSE if the statement is false.

_____ 10. DNA can reveal close evolutionary relationships but cannot help to determine when two organisms began to diverge.

_____ 11. The evolutionary history of an organism is called its phylogeny.

CLASSIFICATION

Interpreting and Applying Concepts

Both cats and dogs are grouped into the same kingdom, phylum, subphylum, class and order. Based on this information, answer the following questions.

1. List some of the common structural features of cats and dogs.

   _____

2. Cats and dogs belong to different families, genera, and species. What can you infer from this? Give some examples to support your answer.

   _____

   _____

Study Figure 14–1, which is a schematic drawing of the five-kingdom classification system, and answer the question.

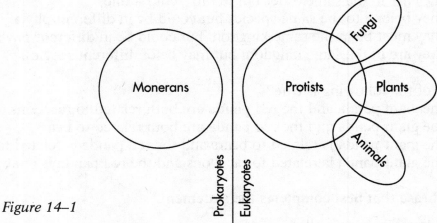

Figure 14–1

3. Explain how this drawing represents some of the problems with the five-kingdom classification system.

   _____

   _____

   _____

   _____

Copyright © by the Glencoe Division of Macmillan/McGraw-Hill School Publishing Company

CLASSIFICATION

Interpreting and Applying Concepts continued

Answer the following questions in complete sentences.

4. Devise a taxonomic key to distinguish among the following: apple, peach, pear, watermelon, orange, apricot, cherry, banana, grapefruit, nectarine.

_____

_____

_____

_____

5. Classification systems have changed over the years as scientists gain more expertise in methods of grouping organisms. One system classified fungi as plants. Describe why you think fungi are now classified in their own kingdom.

_____

_____

_____

6. In your textbook, *Volvox, Spirogyra*, red algae, and *Ulva* are classified as protists, while in other textbooks they are classified as green plants. How do you think algae should be

classified and why? _____

_____

_____

_____

7. The red wolf, and inhabitant of the southeastern United States, was close to extinction in 1975 because of hunting and human encroachment on its habitat. Recently, some scientists have argued that the red wolf is a cross between the coyote and the gray wolf and so should not be placed on the endangered list. How could one demonstrate that the red wolf is a cross and not an endangered species?

_____

_____

_____

_____

_____

_____

CLASSIFICATION

Interpreting and Applying Concepts   continued

8. Maintaining species diversity is of great concern to environmentalists. Discuss reasons why more species are becoming endangered (you may cite specific examples) and suggest possible solutions for solving the problems.

   _____

   _____

   _____

   _____

   _____

   _____

   _____

   _____

   _____

   _____

   _____

   _____

   _____

9. A researcher reported the data below about the number of species in different orders of insects. Make a horizontal bar graph for the data.
   Aptergota (wingless insects) 3500
   Hemiptera (bugs) 60 000
   Othoptera (grasshoppers, locusts, crickets) 20 000
   Coleoptera (beetles) 350 000
   Lepidoptera (butterflies and moths) 120 000
   Hymenoptera (wasps, ants, bees) 100 000
   Diptera (two-winged flies) 120 000

Copyright © by the Glencoe Division of Macmillan/McGraw-Hill School Publishing Company

Name _____ Date _____ Class _____

VIRUSES AND MICROORGANISMS

Understanding Concepts

In the space at the left, write the letter of the word or phrase that best completes the statement or answers the question.

_____ 1. All of the following are diseases caused by viruses EXCEPT
     a. AIDS.   **c.** *Pneumococcus* pneumonia.
     b. measles.   **d.** the flu.

_____ 2. Protozoans are classified on the basis of
     a. type of nutrition.   **c.** method of locomotion.
     b. type of environment.   **d.** method of reproduction.

_____ 3. Which of the following is a common disease-causing organism in Africa?
     a. paramecium   **b.** *Plasmodium*   **c.** radiolarians   **d.** slime mold

_____ 4. Which of the following groups of organisms contains a eukaryotic organism that has been useful in recombinant DNA research?
     a. bacteria   **b.** algae   **c.** fungi   **d.** protozoa

Write the word or phrase that best completes the statement.

5. Viruses are unique in that they can carry on but one life function, _____ .

6. Some RNA viruses contain a specific enzyme, called _____ , that allows them to make DNA from their RNA.

7. Methane-producing bacteria obtain energy by the process of _____ .

8. Bacteria that are important in the recycling of nutrients in organic wastes or dead organisms are called _____ .

9. Fungi are classified by the way in which they _____ .

In the space at the left, write TRUE if the statement is true or FALSE if the statement is false.

_____ 10. Bacteria have flagella that have a structure and function identical to those found in protozoans.

_____ 11. The pigments in blue-green bacteria are found in chloroplasts, where photosynthesis takes place.

_____ 12. Brown algae are complex marine protists important as a commercial source of iodine.

_____ 13. The only mechanism used by fungi to obtain nutrients is saprophytic.

Copyright © by the Glencoe Division of Macmillan/McGraw-Hill School Publishing Company

VIRUSES AND MICROORGANISMS

Interpreting and Applying Concepts

Answer the following questions in complete sentences.

1. A lab technician was culturing euglenoids. She covered the container with foil but cut an opening the size of a quarter in the side of the foil. She then placed a lamp in front of the container. The next day, she removed the foil to observe where the population of euglenoids was densest. Describe what you think she found and give a reason why.

    _____

    _____

    _____

2. Explain why it is difficult to classify the slime molds. _____

    _____

    _____

    _____

3. Bacteria are found on surfaces everywhere--on hay, for example. If hay is placed in pond water containing a small population of paramecia, within one or two days the paramecium population will become fairly large. Explain how this happens.

    _____

    _____

    _____

4. Dogs and cats suffer from viral infections, dogs from canine distemper and cats from feline leukemia, for example. Why do these diseases not spread to humans?

    _____

    _____

    _____

5. Viruses are frequently grown in developing chick embryos. Viruses produced in this way are used to make vaccines for diseases such as smallpox, influenza, and yellow fever. Why are chick embryos useful for the culture of viruses?

    _____

    _____

    _____

   Copyright © by the Glencoe Division of Macmillan/McGraw-Hill School Publishing Company

VIRUSES AND MICROORGANISMS

Interpreting and Applying Concepts continued

6. Viral DNA can be altered and new DNA can be spliced into the DNA of the virus. This is one technique used in recombinant DNA to transfer DNA into host cells. Explain why a virus would be a good way to transport new genes into host cells.

_____

_____

_____

_____

7. In what ways are fungi important to humans? _____

_____

_____

_____

8. In many tropical countries, malaria is a common and often fatal disease. Malaria is caused by a type of protozoan transmitted by the female *Anopheles* mosquito. Mosquitoes live and breed in stagnant water. Many have developed resistance to insecticides commonly used. Given these facts, how might the spread of malaria be prevented? _____

_____

_____

_____

_____

An investigation was performed to determine the rate of growth of a population of bacteria over a period of 16 hours. The data obtained by sampling the population every four hours is given in Table 15–1. Use the table to answer the questions that follow.

Table 15–1

Time In Hours	Number of Bacteria
0	1500
4	14 000
8	76 000
12	58 000
16	20 000

Copyright © by the Glencoe Division of Macmillan/McGraw-Hill School Publishing Company

VIRUSES AND MICROORGANISMS

Interpreting and Applying Concepts   continued

9. According to Table 15–1, when was the rate of bacterial growth greatest?

_____

10. What explanation would you give for the change in population growth after 8 hours?

_____

_____

_____

Read the following paragraph Then answer the questions in complete sentences.

In April 1993, a number of people in Milwaukee, Wisconsin became very sick as a result of contaminated drinking water. They contracted an influenza-like disease. Residents were told to boil water, and schools shut off drinking fountains. The disease, cryptosporidiosis, was caused by a parasite, *Cryptosporidium*, that is resistant to chlorine. The parasite produces egg cells that hatch only when they are in the digestive tract of a host. There, they produce a poison that causes diarrhea and vomiting. The disease is common in countries having poor sanitation and has occurred in places where people drink untreated or improperly treated water. Proper treatment of water supplies will remove the parasite.

11. How do you think the parasite might have gotten into the water supply in Milwaukee?

_____

_____

_____

_____

_____

_____

12. Why might it be difficult to inspect water samples in order to determine the presence of *Cryptosporidium*?

_____

Copyright © by the Glencoe Division of Macmillan/McGraw-Hill School Publishing Company

CHAPTER 16 & ASSESSMENT

PLANT ADAPTATIONS

Understanding Concepts

In the space at the left, write the letter of the word or phrase that best completes the statement or answers the question.

_____ 1. All modern plants are probably descendants of
     a. fungi.   **b.** green algae.   **c.** club mosses.   **d.** horsetails.

_____ 2. Protists include
     a. ferns.   **b.** gymnosperms.   **c.** angiosperms.   **d.** green algae.

_____ 3. An adaptation of land plants that inhibits water loss is the
     a. cell wall.   **b.** chloroplast.   **c.** cuticle.   **d.** vascular tissue.

_____ 4. During photosynthesis, plants
     a. take in oxygen and carbon dioxide.
     b. take in oxygen and give off carbon dioxide.
     c. give off oxygen and carbon dioxide.
     d. take in carbon dioxide and give off oxygen.

_____ 5. Rhizoids are structures analogous to
     a. stems.   **b.** roots.   **c.** leaves.   **d.** chloroplasts.

_____ 6. The sporophyte generation of plants consists of
     a. n cells.   **b.** $2n$ cells.   **c.** $3n$ cells.   **d.** $4n$ cells.

_____ 7. Sporangia are adaptive characteristics of
     a. ferns.   **b.** green algae.   **c.** gymnosperms.   **d.** angiosperms.

_____ 8. Seeds enclosed in a fruit are adaptations of
     a. angiosperms.   **c.** nonvascular plants.
     b. gymnosperms.   **d.** green algae.

_____ 9. The plants best adapted to survival in harsh, dry climates are
     a. green algae.   **b.** ferns.   **c.** liverworts.   **d.** gymnosperms.

In the space at the left, write TRUE if the statement is true or FALSE if the statement is false.

_____ 10. In flowering plants, the male reproductive organs are the stigmas.

_____ 11. Tree bark is an example of a plant adaptation to reduction of water loss.

_____ 12. Nonvascular plants are well-adapted for growth in dry environments.

_____ 13. Gymnosperms have reproductive structures in needle-shaped leaves.

_____ 14. In a moss, the cells of the gametophyte generations are haploid.

PLANT ADAPTATIONS

Interpreting and Applying Concepts

All living things possess characteristics that form the basis of their classification. The table below shows certain characteristics of green algae, nonvascular plants, and vascular plants. Based on the data in Table 16–1, answer the questions that follow.

Table 16–1

Characteristics of Green Algae, Nonvascular Plants, and Vascular Plants	
Organism	**Characteristics**
Green Algae	1. Manufacture food through photosynthesis 2. Have cell walls containing cellulose 3. Make starch 4. Limited to watery or moist environment
Nonvascular Plant	1–4 above plus: 5. Possesses a cuticle that inhibits water loss 6. Takes in carbon dioxide from air
Vascular Plant	1–3 and 5–6 above, plus: 7. Possesses structures that provide support 8. Possesses true roots, stems, and leaves 9. Usually grows very tall 10. Can live in a wide variety of environments

1. What data in Table 16–1 would support the hypothesis that plants evolved from green algae?

   _____

   _____

2. If you discovered a new plant species in a desert, what inference could you make concerning its probable classification? Why?

   _____

   _____

   _____

3. Suppose you have a specimen of a single cell from an organism that you have not observed. You examine the cell under a microscope. You discover that the cell has a rigid cell wall and chlorophyll. What conclusions could you reach, or not reach, concerning the classification of the organism from which the cell came? Why?

   _____

   _____

PLANT ADAPTATIONS

Interpreting and Applying Concepts   continued

4. A few days later, you obtain a larger sample of the organism. You find the specimen is covered with a cuticle. What can you now infer about the classification of the organism? Why?

_____

_____

5. You really want to pin down the classification of this mysterious organism. You go out into the field in search of the whole plant. Unfortunately all you find is a clump of soil in which you discover the organism's root. This clue allows you to classify the organism into a single group; what group and why?

_____

While on a field trip, you collect an assortment of plant parts. You label each part with a code number. The parts and their numbers are shown in Figure 16–1. When you get back to your lab, you complete the classification table, Table 16–2. After completing the table, answer the following questions.

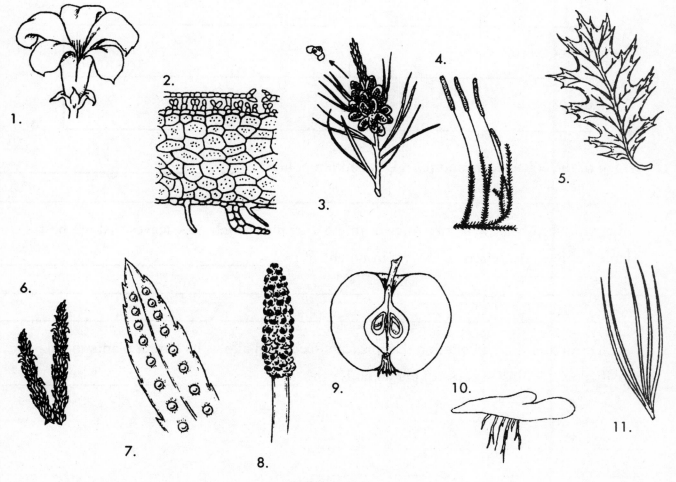

Figure 16–1

PLANT ADAPTATIONS

Interpreting and Applying Concepts   continued

Table 16–2

Classification Table		
Organism	**Code Number**	**Description/Identification**
Moss		
Liverwort		
Club moss		
Horsetail		
Fern		
Gymnosperm		
Angiosperm		

1. Which of the collected structures might contain pollen? _____

    _____

2. Among the seed-bearing plants, infer from the drawings which have leaves that are best
    adapted to a windy, relatively dry environment. Why? _____

    _____

    _____

3. Which structure is an adaptation that protects seeds and allows its parent plants to be
    classified as the most successful plants on Earth? _____

Copyright © by the Glencoe Division of Macmillan/McGraw-Hill School Publishing Company

CHAPTER 17 ✿ ASSESSMENT

ANIMAL ADAPTATIONS

Understanding Concepts

In the space at the left, write TRUE if the statement is true or FALSE if the statement is false.

_____ 1. Sponges are classified according to the structure and composition of spicules.

_____ 2. In sponges, reproduction from body fragments results in genetic variability.

_____ 3. Tissues in cnidarians are evidence of an evolutionary advance over sponges.

_____ 4. In general, bilateral symmetry is typical of the most primitive animals.

_____ 5. Flatworms are the first animals in the evolutionary tree to possess two body openings.

_____ 6. Flatworms reproduce by cross-fertilization, which ensures genetic variability.

_____ 7. Each segment in a segmented worm contains different structures.

_____ 8. Segmented worms possess a primitive brain but no nerve cord.

_____ 9. Some mollusks do not have shells.

_____ 10. The appendages of an arthropod allow for a variety of movements.

_____ 11. Internal fertilization is an adaptation of land-dwelling animals.

Answer the following questions.

12. In what way is the decentralization of an echinoderm's nervous system an adaptive advantage?

_____

13. What led taxonomists to separate the hemichordates and chordates into two different phyla?

_____

14. What is the major characteristic that distinguishes organisms in the subphyla that contain tunicates and lancelets from vertebrates? _____

_____

15. What distinguishes the two major groups of vertebrates? _____

_____

16. What is the main difference between the reproductive process of amphibians and those of reptiles, birds, and mammals? _____

_____

ANIMAL ADAPTATIONS

Interpreting and Applying Concepts

Figure 17–1 presents data for the eight vertebrate classes. Based on the chart, answer the following questions.

Figure 17–1

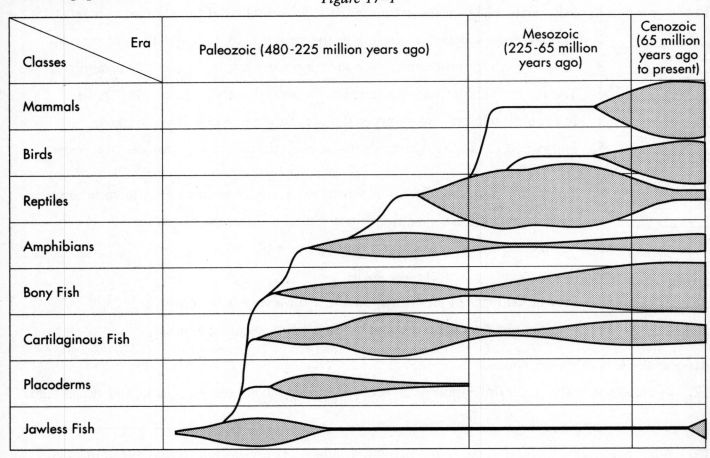

Classes \ Era	Paleozoic (480-225 million years ago)	Mesozoic (225-65 million years ago)	Cenozoic (65 million years ago to present)
Mammals			
Birds			
Reptiles			
Amphibians			
Bony Fish			
Cartilaginous Fish			
Placoderms			
Jawless Fish			

1. Suggest an appropriate title for the chart.

    _____

2. A distinguishing characteristic of birds and mammals is that they are endothermic. Express a hypothesis and the reasoning behind it, to support the possibility that extinct members of another class were endothermic.

    _____

    _____

    _____

    _____

    Copyright © by the Glencoe Division of Macmillan/McGraw-Hill School Publishing Company

ANIMAL ADAPTATIONS

Interpreting and Applying Concepts   continued

3. The chart implies that amphibians evolved after fish but before reptiles. What characteristic of frogs supports this implication? How does it do so?

_____

_____

4. How can the data in the chart be interpreted to reveal which class of reptiles is extinct today?

_____

5. Interpret the chart to determine which class gave rise to two other classes. What are the classes and which gave rise to which?

_____

6. What era could be considered the age of reptiles? Use the given data to support your answer.

_____

_____

Scientists are involved in a controversy involving the positioning of two fish, the coelacanth and the lungfish, on the vertebrate family tree. Scientists have proposed two conflicting hypotheses about these fish. One suggests that the coelacanths are more closely related to tetrapods than to lungfish. The other suggests that the opposite is true. Figure 17–2 shows the vertebrate family tree with two branches unoccupied. Table 17–1 reveals data concerning the coelacanth, lungfish, and tetrapods. Based on the figure and the table, answer the questions that follow.

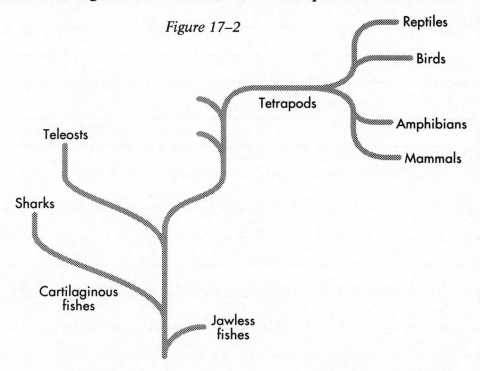

Figure 17–2

ANIMAL ADAPTATIONS

Interpreting and Applying Concepts   continued

Table 17–1

Characteristics of Coelacanths, Lungfish, and Tetrapods		
Coelacanth	**Lungfish**	**Tetrapod**
gills, degenerate lungs	gills and air-breathing lunglike structures	air-breathing lungs in all adults
limbs are leglike	limbs are leglike	all classes have members that have legs
hemoglobin part similar to tadpole's	————	————
————	mitochondrial DNA more similar to frog's than coelacanth's mitochondrial DNA	————
————	ear like shark's	————
control of eyes more like that of tetrapods	control of eyes less like that of tetrapods	————

7. Based on the family tree and the datum regarding ear structure, which hypothesis is supported? Why? _____

_____

8. Based on the family tree and the data regarding mitochondrial DNA, which hypothesis is supported? Why? _____

_____

9. Why does the respiratory anatomy of coelacanths and lungfish support or not support either hypothesis? _____

_____

10. Based on the family tree and the data regarding hemoglobin, which hypothesis is supported? Why? _____

_____

_____

11. Where on the family tree do you believe the coelacanth and lungfish should be? Explain.

_____

_____

Copyright © by the Glencoe Division of Macmillan/McGraw-Hill School Publishing Company

CHAPTER 18 🐚 ASSESSMENT
REPRODUCTION

Understanding Concepts

In the space at the left, write the letter of the word or phrase that best completes the statement or answers the question.

_____ 1. Bulbs are food storage organs that aid a plant in
    a. spreading out over a large area.   **c.** supplying food for animals and people.
    b. surviving over winter.   **d.** resisting infection.

_____ 2. Vegetative reproduction in a creeping stem differs from the growth of an ordinary stem because after a creeper is rooted, it
    a. separates from the parent plant   **c.** dies off.
    and grows independently.
    b. grows faster.   **d.** produces seeds.

_____ 3. Fragmentation is a kind of asexual reproduction for starfish because they can
    a. break open a clam.   **c.** regenerate all the missing parts of their body from a single small part.
    b. escape by leaving a limb behind.   **d.** break into many parts.

_____ 4. A gardener grows a new plant from a cutting by means of
    a. budding.   **b.** corms.   **c.** tubers.   **d.** regeneration.

_____ 5. An animal produced by parthenogenesis develops from
    a. a fertilized egg.   **c.** a regenerated part.
    b. an unfertilized egg.   **d.** diploid spore.

In the space at the left, write the letter of the phrase from Column B that best matches the term in Column A.

Column A	Column B
_____ 6. budding	**a.** transfer of pollen from anther to stigma
_____ 7. pollination	**b.** genetic material transferred by cell-to-cell contact
_____ 8. estrus	**c.** method of reproduction in *Hydra*
_____ 9. fragmentation	**d.** individual in which both kinds of sex organs develop
_____ 10. ovulation	**e.** energy source in some new, developing plants
_____ 11. gametes	**f.** period of readiness for mating
_____ 12. conjugation	**g.** method of reproduction in planarian
_____ 13. ovules	**h.** become seeds after fertilization
_____ 14. endosperm	**i.** sex cells
_____ 15. hermaphrodite	**j.** release of an egg from a follicle

Copyright © by the Glencoe Division of Macmillan/McGraw-Hill School Publishing Company

REPRODUCTION

Interpreting and Applying Concepts

Animals that reproduce by means of external fertilization have various adaptations to ensure fertilization of the eggs. The stickleback, a small fish, has been studied to find out what factors affect its reproductive success. Stickleback males build a nest and try to induce females to lay their eggs in the nest. Two factors seem to influence a female's choice of a male: the male's mating coloration and the location of the site selected by the male to build his nest. Study the graphs in Figure 18–1 and answer the following questions in complete sentences.

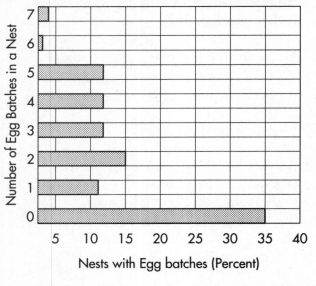

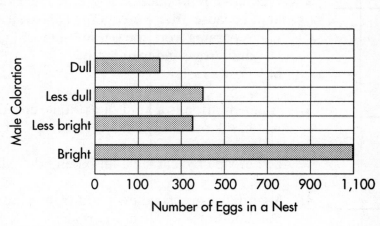

Figure 18–1

1. What percentage of males' nests received no batches of eggs? _____

2. What percentage of the nests received 4 or more egg batches? _____

3. What could you conclude from the graph about the success of some males in attracting

   females? _____

   _____

4. What does the male coloration graph seem to show about the reproductive success of brightly colored males over males with less bright coloration?

   _____

5. In the male coloration graph, less dull sticklebacks attracted females to lay more eggs in their nests than were laid in the nests of less bright sticklebacks. What factors besides color may be

   affecting the data? _____

   _____

   _____

   _____

Copyright © by the Glencoe Division of Macmillan/McGraw-Hill School Publishing Company

REPRODUCTION

Interpreting and Applying Concepts   continued

6. There are other factors that affect the mating behavior of sticklebacks. Males often eat the eggs fertilized by their competitors, and females often eat eggs laid by other females. How would these behaviors affect the data on the graphs? _____

_____

In Australia, the aquatic snail *Potamopyrgus antipodarum* **was found to have individuals that reproduced asexually and others that reproduced sexually. These snails are commonly infected with a parasitic worm. To find out if a snail's method of reproduction affects the degree to which it becomes infected with the worms, snails and worms were collected from two lakes, A and B, located 3050 m apart. Snail populations from both lakes had individuals that reproduced sexually and others that reproduced asexually. Use the diagrams in Figure 18–2 to answer the questions.**

Figure 18–2

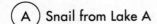

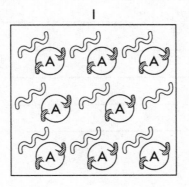

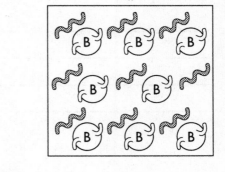

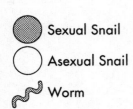

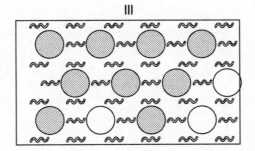

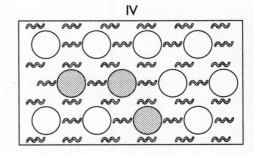

7. Snails from Lake A were placed in container I with worms from both lakes. What happened to the snails? _____

8. When snails from Lake B were placed in container II with worms from both lakes, what happened to the snails? _____

9. A number of snails that reproduce sexually were placed in container III with an equal number of snails that reproduce asexually and a large number of worms. What happened to the asexual snails after a prolonged period of time?

_____

Copyright © by the Glencoe Division of Macmillan/McGraw-Hill School Publishing Company

REPRODUCTION

Interpreting and Applying Concepts   continued

10. In container IV, equal numbers of sexual snails and asexual snails were placed with a small number of worms. What happened to the asexual snails after a prolonged period of time?

_____

11. Hypothesize why these changes in the kind of reproductive behavior were selected for.

_____

_____

_____

_____

12. Why might clones be easy prey for parasites? _____

_____

_____

_____

_____

   Copyright © by the Glencoe Division of Macmillan/McGraw-Hill School Publishing Company

CHAPTER 19 🐛 ASSESSMENT

DEVELOPMENT

Understanding Concepts

In the space at the left, write the letter of the word or phrase that best completes the statement or answers the question.

_____ 1. Embryos of both plants and animals are similar in
    a. the timing of early cell division and growth of the embryo.
    b. having a food source for the embryo.
    c. development that results from different rates and patterns of cell division.
    d. development that results from cell movement.

_____ 2. Morphogenesis occurs when the
    a. egg is fertilized.
    b. zygote begins to divide.
    c. outer cells move inward and the first evidence of body systems appears.
    d. unfertilized egg is produced.

_____ 3. During the third trimester of human gestation, the fetus increases
    a. its mass.
    b. the number of body systems.
    c. the complexity of its body.
    d. its endoderm tissue.

_____ 4. Unlike animals, many plants can develop new structures throughout their lifetime because
    a. they never stop being embryos.
    b. if they stop developing, they die.
    c. they have regions of cell division, called meristems, in many parts of the plant.
    d. they make sugars that provide energy.

Write the word or phrase that best completes the statement.

5. The group of cells produced as a result of cleavage is called the _____ .

6. The _____ is the beginning of the nervous system in the developing embryo.

7. A seed develops into a new plant during _____ , which occurs if conditions are favorable for growth.

8. The _____ becomes the main food source for the developing plant embryo after the endosperm is depleted.

9. When one part of an embryo influences the development of another part, _____ occurs.

10. The stage of development represented by a tadpole is called a(n) _____ .

11. The _____ is tissue formed from both the mother's uterus and the embryo itself, and is the site of nutrient and waste exchange.

Copyright © by the Glencoe Division of Macmillan/McGraw-Hill School Publishing Company

DEVELOPMENT

Interpreting and Applying Concepts

Not all frogs develop in the usual way from egg to tadpole to four-legged adult. In fact, at least fourteen of the twenty-four known families of frogs show variations in the way they develop. These variations fall into three categories: (1) direct development, (2) transport of eggs or tadpoles by one of their parents, (3) development in foam nests. Direct development means developing from eggs to four-legged froglets without a tadpole stage. Frogs, male or female, that transport eggs or tadpoles may do so on their back, in their vocal sacs, or even inside their stomach! For those eggs covered by foam when they are laid, the outer layer of the foam hardens, providing greater protection. Study the diagram in Figure 19–1 which shows data for some families of frogs. Then answer the following questions.

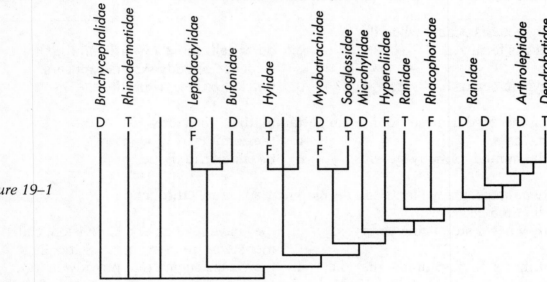

Figure 19–1

D   Direct development
T   Transported by a parent during development
F   Development in a foam nest
    Lines show relationships

1. How many of the families of frogs shown in the diagram have species that develop directly from eggs to froglets without the tadpole stage?

   _____

2. What selective advantage might there be in direct development for frogs?

   _____

   _____

   _____

3. How many of the frog families shown have more than one unusual strategy for development?

   _____

    Copyright © by the Glencoe Division of Macmillan/McGraw-Hill School Publishing Company

DEVELOPMENT

Interpreting and Applying Concepts   continued

4. What factors may select for one type of development over another?

_____

_____

5. When would protective foam around the eggs provide a selective advantage?

_____

_____

_____

6. What advantage is there to one of the parents carrying the eggs or tadpoles on or in its body?

_____

_____

7. Use Figure 19–1 to tell whether or not family relationships affect the kinds of development strategies a family engages in. _____

_____

_____

_____

_____

_____

Read the following paragraph and answer the questions in complete sentences.

Scientists have tried to understand how an animal with many complex systems develops from a single-celled zygote. In the late nineteenth century, Wilhelm Roux, a German embryologist, studied the development of fertilized frog eggs. He investigated what would happen if he killed one cell of a two-celled embryo with a hot needle. Roux found that the live cell developed as half an embryo. Roux concluded that the abnormal development occurred because half of the material needed to determine a complete embryo was missing after he killed one cell. Hans Driesch performed a somewhat similar experiment with the two-celled stage of sea-urchin embryos. Instead of killing one of the cells, Driesch separated the two cells and observed what happened. Both cells continued to develop and to form two complete embryos, smaller than usual, but otherwise normal.

8. What was the purpose of the experiments carried out by Roux and Driesch?

_____

_____

DEVELOPMENT

Interpreting and Applying Concepts   continued

9. Most of the scientists in the eighteenth and early nineteenth centuries believed that all the embryo parts were complete (or completely determined) at fertilization rather than developing later. What do you think was the impact of each of the two experiments on this theory?

_____

_____

_____

_____

_____

10. Perhaps you concluded that the difference in the results of the two experiments occurred because Roux used frog eggs and Driesch used sea-urchin eggs. How could you prove that the kinds of eggs used were not the cause of the different results?

_____

_____

_____

_____

_____

_____

11. Is it fair to say that at the two-cell stage, adjustments can still be made to the developmental

process? Explain. _____

_____

_____

12. Beatrice Mintz investigated to see if one mouse embryo could be formed from two embryos. She removed the membrane around two eight-celled embryos and allowed the two embryos to fuse. After fusion, she placed the new embryo in the uterus of a mouse. A normal mouse was born. What does this investigation prove about early development in mice?

_____

_____

_____

   Copyright © by the Glencoe Division of Macmillan/McGraw-Hill School Publishing Company

CHAPTER 20 ❧ ASSESSMENT
NUTRITION AND DIGESTION

Understanding Concepts

In the space at the left, write the letter of the word or phrase that best completes the statement or answers the question.

_____ 1. Earthworms lack teeth but are able to grind their food into pieces in the
     a. intestine.   **b.** gizzard.   **c.** stomach.   **d.** mouth.

_____ 2. An animal that digests its food in a gastrovascular cavity provides a simple example of
     a. extracellular digestion.   **c.** digestion without enzymes.
     b. intracellular digestion.   **d.** a digestive tract with two openings.

_____ 3. The end product of protein digestion is
     a. bile.   **b.** maltose.   **c.** amino acids.   **d.** monosaccharides.

_____ 4. One end product of the digestion of fats is
     a. lipase.   **b.** cholesterol.   **c.** glucose.   **d.** glycerol.

_____ 5. Examples of heterotrophs that carry out intracellular digestion are
     a. *Rhizopus* and mushrooms.   **c.** *Hydra* and the earthworm.
     b. Venus's-flytrap and the pitcher plant.   **d.** *Amoeba* and *Paramecium*.

Write the word or phrase that best completes the statement.

6. The digestion of starch begins in the _____ and continues in
the _____ .

7. Protein digestion begins in the _____ and continues in
the _____ .

8. Fats are digested in the _____ .

9. The surface area of the small intestine is increased by _____ in its lining.

10. The _____ converts excess glucose in the blood to glycogen.

In the space at the left, write TRUE if the statement is true or FALSE if the statement is false.

_____ 11. Plants do not require the same kinds of organic substances that heterotrophs
need in order to grow.

_____ 12. Heterotrophs obtain food by eating other organisms.

_____ 13. In filter-feeding mollusks, both intracellular and extracellular digestion occur in
the stomach and intestines.

_____ 14. After digestion, excess amino acids are converted to proteins.

Copyright © by the Glencoe Division of Macmillan/McGraw-Hill School Publishing Company

NUTRITION AND DIGESTION

Interpreting and Applying Concepts

Iron is one of the minerals important to the health of the body. Although many foods contain iron, the body does not absorb the iron from plants as well as it does the iron from animal products. Study Table 20–1, which lists both plant and animal sources of iron. The table shows how much iron a 90-gram portion of each food contains and how much of this iron can be absorbed by the body. Answer the following questions in complete sentences.

Table 20–1 **SOURCES OF IRON**

	Iron Content (mg)	Iron Absorbed (mg)	% of Iron Absorbed
Iceberg lettuce	0.5	0.02	4.0
Spinach (cooked)	2.0	0.04	2.0
Black beans (cooked)	4.3	0.07	1.6
Soybeans (cooked)	5.0	0.35	7.0
Chicken (roasted, no skin)	1.2	0.22	18.3
Ground beef (broiled)	3.0	0.60	20.0
Sirloin steak (broiled)	5.4	1.08	20.0
Calf liver (fried)	12.0	1.80	15.0

1. Which food from plants is the best source of iron? Explain. _____

   _____

2. Explain why lettuce is not a good source of iron even though the percent of iron absorbed is higher than that of some of the other plants. _____

   _____

   _____

3. Which meat source would you choose for its iron content and ability to be absorbed by the body? _____

   _____

4. Use the information about iron from plant and animal sources to explain why many people from poorer regions of the world suffer from an iron deficiency. _____

   _____

5. Vitamin C helps and tea hinders the absorption of iron from plants. Based on this information, what recommendation might you make to a vegetarian friend? _____

   _____

  Copyright © by the Glencoe Division of Macmillan/McGraw-Hill School Publishing Company

NUTRITION AND DIGESTION

Interpreting and Applying Concepts continued

Athletes at an advanced level of physical fitness have different nutritional needs from those at an intermediate level. The athletes at each level have a set Calorie requirement based on their physical activity. They keep track of the food units to be consumed daily, based on the food pyramids in Table 20–2. Answer the following questions in complete sentences.

Table 20–2 **NUTRITIONAL NEEDS FOR ATHLETES**

Food Category (1 Serving = 1 Unit)	Units/Advanced Level		Units/Intermediate Level	
	Female 2200 Cal.	Male 2800 Cal.	Female 1900 Cal.	Male 2500 Cal.
Meat (1 oz)	6	7	5	7
Dairy (1 cup)	3	3	2	2
Fat (1 tsp)	7–8	10–11	6–7	8–9
Fruit (1)	3	4	3	4
Vegetables (leafy – 1 cup; other – 1/2 cup)	4–4.5	4.5–5.5	4–4.5	4.5–5.5
Grain (1 slice bread; 1 cup dry cereal; 1/2 cup cooked cereal, pasta, rice)	9	13	8	12

6. Which food units are reduced for male athletes from the advanced to the intermediate level?

_____

7. How many more additional overall food units does a female athlete at the advanced level have to consume compared to a female athlete at the intermediate level?

_____

8. Some athletes also work out at the beginner level. Suppose the number of Calories needed for the beginner level is decreased by the same amount that the number of Calories at the intermediate level is decreased from the advanced level. How many Calories will the beginner

female and male athletes consume? _____

9. Female athletes at the beginner level will eat 4 food units less, equally divided between the fat and grain categories, than female athletes at the intermediate level. How many units of fats

and grains will beginner female athletes consume? _____

10. If the male athletes at the beginner level need the same number of food units as the female athletes at the advanced level, how many fat and grain units will beginner male athletes

consume? _____

NUTRITION AND DIGESTION

Interpreting and Applying Concepts   continued

The tally for the food units consumed by Jennifer and Jeff during three main meals are listed on the chart below.

	Meat	Dairy	Fat	Fruit	Vegetable	Grain
Jennifer	6	1	7	2	4.5	7
Jeff	7	2	10	2	5.5	9

11. If both Jennifer and Jeff are at the advanced level, what additional food units should they

    choose for snacks? _____

    _____

12. You have heard so much about the importance of reducing fat consumption. Why then are

    Jeff's allowable fat units so high? _____

    _____

    _____

80   CHAPTER 20   ASSESSMENT   Copyright © by the Glencoe Division of Macmillan/McGraw-Hill School Publishing Company

CHAPTER 21 ❧ ASSESSMENT

TRANSPORT

Understanding Concepts

In the space at the left, write the letter of the word or phrase that best completes the statement or answers the question.

_____ 1. The transport system of vascular plants consists of
    a. roots and stems.   **c.** xylem and phloem.
    b. buds and leaves.   **d.** cells and root hairs.

_____ 2. The transport of oxygen in the blood is the function of the
    a. red blood cells. **b.** white blood cells. **c.** antibodies. **d.** antigens.

_____ 3. Cell fragments that help repair cuts by clotting the blood are
    a. red blood cells. **b.** white blood cells. **c.** platelets. **d.** antigens.

Write the word or phrase that best completes the statement.

4. After _____ cells in plants die, they form hollow tubes that transport water to the leaves and other parts of the plant.

5. The process by which water is lost through the pores in stomata is called _____.

6. Oxygen, carbon dioxide, and nitrogenous wastes pass through the capillary wall by the process of _____ .

7. The largest blood vessel in the body is the _____ .

In the space at the left, write the letter of the phrase from Column B that best matches the term in Column A.

Column A	Column B
_____ 8. root hair	**a.** proteins on the membranes of red blood cells
_____ 9. vein	**b.** place where an electrical impulse causes the ventricles to contract
_____ 10. lymph node	**c.** site where gases and nutrients are exchanged between the body cells and the blood
_____ 11. white blood cell	**d.** carries blood to the heart
_____ 12. antigens	**e.** living phloem cells that transport sucrose
_____ 13. sieve tube	**f.** place where most bacteria in the body are destroyed
_____ 14. artery	**g.** absorbs water and dissolved ions from the film of moisture around soil particles
_____ 15. capillary	**h.** carries blood away from the heart
_____ 16. type O blood	**i.** cell that defends against viruses, microorganisms, and parasites
_____ 17. atrioventricular node	**j.** universal donor

Copyright © by the Glencoe Division of Macmillan/McGraw-Hill School Publishing Company

TRANSPORT

Interpreting and Applying Concepts

To demonstrate how sucrose, a simple sugar, is transported in a plant, researchers girdled a tree. Girdling involves removing a ring of bark around the tree. The girdle cuts through and destroys the phloem tissue. It has no effect on the xylem, which is closer to the center of the tree stem. Study Figure 21–1 and answer the following questions.

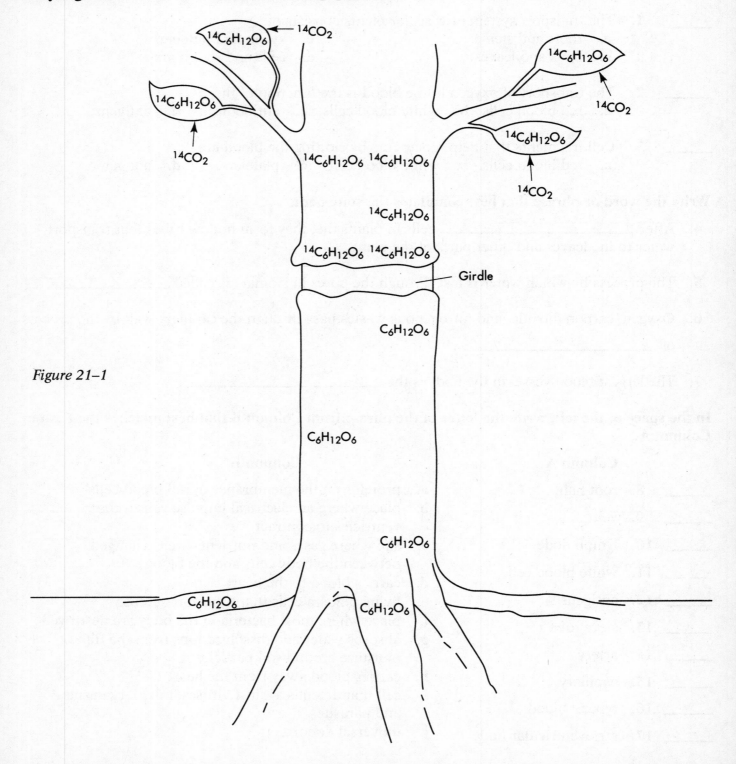

Figure 21–1

TRANSPORT

Interpreting and Applying Concepts   continued

1. The researchers supplied carbon dioxide labeled with radioactive carbon (carbon–14) to the leaves of the girdled tree. What happened to the radioactive carbon in the leaves?

   _____

2. Where is the radioactive carbon found in the tree stem?

   _____

3. Study Figure 21–2 to see if the radioactive carbon reached the roots.

   _____

4. What can you deduce about the role of phloem in the transport of sucrose from the leaves? Why is there a bulge in the stem above the girdle?

   _____

   _____

   _____

   _____

5. How could the experimenters be certain that xylem tubes had not transported the radioactive

   sugars? _____

   _____

   _____

6. How do you know that the sucrose in the lower stem and roots was produced in the leaves

   before the girdling took place? _____

   _____

   _____

   _____

   _____

TRANSPORT

Interpreting and Applying Concepts   continued

Blood pressure changes as the blood flows through the body. A blood pressure reading includes the *systolic pressure,* **the pressure during the contraction of the ventricles, over the** *diastolic pressure,* **the pressure during the relaxation of the ventricles. The highest blood pressure occurs in the ventricles. Use the information about the changes in blood pressure (Figure 21–2) and the velocities of blood flow to answer the following questions.**

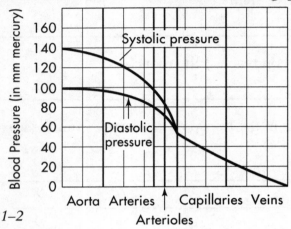

VELOCITIES OF BLOOD FLOW

Vessels	Velocity (cm/sec)
Aorta	40
Arteries	40–10
Arterioles	10–0.1
Capillaries	0.1
Venules	0.3
Veins	0.3–5.0
Vena Cavae	5–20

Figure 21–2

7. What is the patient's blood pressure as shown in the graph when the blood enters the aorta?

_____

8. What happens to the blood pressure as the blood enters the arteries? _____

_____

9. What is the blood pressure in the arterioles? _____

10. What change do you notice in the blood pressure of the capillaries? _____

_____

11. Relate the pressures to the velocities of the blood as it flows through the body.

_____

_____

_____

_____

12. Why is it a selective advantage that the blood slows down in the capillaries? _____

_____

_____

   Copyright © by the Glencoe Division of Macmillan/McGraw-Hill School Publishing Company

CHAPTER 22 ❧ ASSESSMENT

GAS EXCHANGE AND EXCRETION

Understanding Concepts

In the space at the left, write the letter of the word or phrase that best completes the statement or answers the question.

_____ 1. The kangaroo rat does not drink water because
    a. water is harmful to its body.
    b. it does not need water to live.
    c. its kidneys resorb almost all the water that passes through them.
    d. its food is mostly liquid.

_____ 2. The excess salt taken in by marine fish is removed by
    a. active transport across the gills.
    b. the kidneys.
    c. being absorbed by the cells.
    d. nephridia.

_____ 3. A paramecium maintains osmotic balance by removing excess water by means of
    a. cilia.
    b. excretory ducts.
    c. flame cells.
    d. contractile vacuoles.

_____ 4. In humans, most of the water flowing through the kidneys
    a. stays in the glomerulus.
    b. is resorbed by the kidneys.
    c. passes out in the urine.
    d. remains in the stomach.

_____ 5. In fish, oxygen-rich water
    a. leaves through the mouth.
    b. is expelled through the gills.
    c. passes over the gills, where oxygen is removed.
    d. is transported through the body.

In the space at the left, write the letter of the phrase from Column B that best matches the term in Column A.

Column A		Column B
_____ 6. vasopressin	**a.**	tube through which urine is excreted
_____ 7. ureter	**b.**	muscular storage sac for urine
_____ 8. glomerulus	**c.**	removal of nitrogenous wastes
_____ 9. urinary bladder	**d.**	tube from the kidney that conveys urine to the bladder
_____ 10. flame cell	**e.**	part of the excretory system in a planarian
_____ 11. urethra	**f.**	mass of capillaries in a nephron
_____ 12. Malpighian tubule	**g.**	excretory organ in an earthworm
_____ 13. nephron	**h.**	excretory organ in a grasshopper
_____ 14. excretion	**i.**	hormone that controls the amount of water excreted
_____ 15. nephridium	**j.**	tiny excretory unit in the kidney

GAS EXCHANGE AND EXCRETION

Interpreting and Applying Concepts

At elevations of about 3500 meters above sea level, the decreased amount of oxygen available begins to affect the human respiratory system. People experience the symptoms of altitude sickness, including dizziness and blurred vision. The graphs in Figure 22–1 show the results of some studies conducted to determine the effects of high altitudes on humans. Study the graphs and answer the following questions in complete sentences.

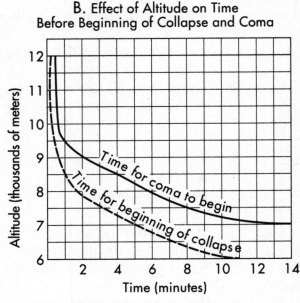

A. Effect of Breathing Air or Oxygen on Saturation of Oxygen in Arteries

Figure 22–1

B. Effect of Altitude on Time Before Beginning of Collapse and Coma

1. Refer to Figure 22–1A. Up to what altitude does the oxygen in the arteries of a person breathing pure oxygen remain close to 100 percent saturation? How does this compare to someone breathing air? _____

_____

_____

2. Refer to Figure 22–1A. The lowest level of arterial oxygen at which a person can remain alive for more than a few hours is 50 percent. What is the highest altitude at which someone breathing pure oxygen can stay alive? _____

_____

3. Refer to Figure 22–1A. Some people experience mild altitude sickness when breathing air at 3000 m. At what saturation level is the oxygen in their arteries?

_____

4. Refer to Figure 22–1A. There is an altitude above which a person cannot survive breathing air. There is a different altitude above which a person cannot survive breathing pure oxygen. How do these two altitudes compare? _____

_____

Copyright © by the Glencoe Division of Macmillan/McGraw-Hill School Publishing Company

GAS EXCHANGE AND EXCRETION

Interpreting and Applying Concepts continued

5. Refer to Figure 22–1B. The great danger of high altitudes is the impairment of a person's ability to respond when deprived of sufficient oxygen. "Collapse" means extreme weakness and mental haziness. "Coma" means unconsciousness. Find out how long a mountain climber has to replace an oxygen mask ripped off during a fall while almost at the peak of

Mount Everest (8900 m). _____

_____

6. A person who stays at a high altitude for several weeks usually will recover from altitude sickness. The body adapts in the following ways: the bone marrow produces greater numbers of red blood cells than usual, the respiratory center of the brain is stimulated, and the size of the alveoli may increase. Explain how each of these changes helps the person overcome the

effects of the high-altitude conditions. _____

_____

_____

_____

A water environment contains certain ions (mostly sodium, NA$^+$, chlorine, Cl$^-$, and magnesium, Mg^{++}). The ions combine to form salts. Often the concentrations of these ions in the bodies of the animals that live in water differ from the concentrations of ions in the water in which they live. If an animal's ionic concentrations are higher than its surroundings, it tends to gain water and lose ions. If an animal's ionic concentrations are lower than its surroundings, it loses water and gains ions. Study the graph in Figure 22–2, which compares the relative concentrations of ions in animals that live in water with the concentrations of the ions in seawater and in fresh water. Then answer the following questions in complete sentences.

Figure 22–2

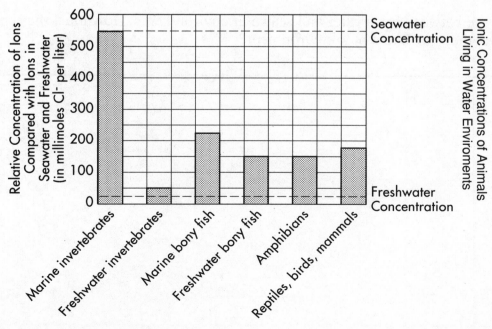

GAS EXCHANGE AND EXCRETION

Interpreting and Applying Concepts   continued

7. Which animals have body fluids with an ionic content closest to that of seawater?

_____

8. Which animals have ionic concentrations similar to that of fresh water?

_____

9. How might you explain why the ionic concentrations of marine bony fish are closer to that of fresh water than to the seawater in which it lives?

_____

10. Compare the ion concentrations in freshwater bony fish with that of their environment. What must these fish do to survive in their environment?

_____

_____

_____

_____

11. Compare the ion concentrations in marine bony fish with that of their environment. How can these fish survive in their environment?

_____

_____

_____

12. Salmon move between freshwater and seawater environments. How must they change their habits when they move to a new environment?

_____

Copyright © by the Glencoe Division of Macmillan/McGraw-Hill School Publishing Company

Name _____ Date _____ Class _____

IMMUNE SYSTEM

Understanding Concepts

In the space at the left, write the letter of the word or phrase that best completes the statement or answers the question.

_____ 1. The bacteria that cause strep throat are
    a. endotoxins.   **b.** exotoxins.   **c.** pathogens.   **d.** macrophages.

_____ 2. A specific defense mechanism that guards your body from disease-causing microorganisms is
    a. release of histamines.   **c.** saliva.
    b. hairs in the nasal passage.   **d.** unbroken skin.

_____ 3. White blood cells that are produced in the bone marrow but mature in the thymus gland are
    a. T cells.   **b.** B cells.   **c.** receptors.   **d.** antigens.

_____ 4. Which of the following are not part of the inflammatory response?
    a. fever   **c.** production of large numbers of white blood cells
    b. swelling   **d.** acids and enzymes in your body

Write the word or phrase that best completes the statement.

5. Poisonous chemicals secreted by living bacteria are known as _____ .

6. _____ are white blood cells that recognize antigens and begin to destroy specific pathogens.

7. A(n) _____ defends against body cells in which viruses are reproducing.

8. A(n) _____ is an abnormal response of the immune system to part of a person's own body.

9. A person who receives antibodies made by another animal acquires a type of protection against disease called _____ .

10. _____ are proteins in T cells that recognize certain self proteins.

11. The lymph vessels, lymph nodes, spleen, tonsils, bone marrow, and thymus gland are all part of the _____ .

12. _____ are white blood cells that engulf and digest pathogens.

Copyright © by the Glencoe Division of Macmillan/McGraw-Hill School Publishing Company

IMMUNE SYSTEM

Interpreting and Applying Concepts

Inbred mice in which brothers and sisters are mated can produce offspring that are homozygous for a certain trait. Continued crossing for about 20 generations can produce mice in which all the traits are exactly the same. Two different types of purebred mice—A and B—were produced in this way. Then A mice were crossed with B mice; the offspring were called A × B mice. The parents involved in the cross were then given radiation treatment to destroy all their lymphocytes. The irradiated parent mice immediately received donations of lymphocytes from the bone marrow or spleen of their A × B offspring. The lymphocytes from an offspring's bone marrow were immature at the time of donation, that is, they had not yet learned to distinguish between self and non-self proteins. Lymphocytes from spleen cells were already mature when donated; they had learned to distinguish between self and non-self before donation. The lymphocyte donations caused the A and B mice to have lymphocytes that were exactly like the donor's, while the mice's other tissues were as they had been before the donation. Table 23–1 shows what happened when the irradiated mice with donor lymphocytes were vaccinated with A cells infected with vaccinia virus (A-vaccinia) or B cells infected with vaccinia virus (B-vaccinia). The vaccinations were carried out for a total of four experiments, and the results were recorded in the table. (Note: The vaccinia virus causes cowpox. This is the same virus used by Edward Jenner almost 200 years ago to produce the first vaccinations against smallpox.) Using the information in Table 23–1, answer the following questions.

Table 23–1

Experiment	Donor Cells	Recipient	Response to infected cells	
			A-vaccinia	B-vaccinia
1	A × B (from bone marrow)	A	+	–
2	A × B (from bone marrow)	B	–	+
3	A × B (from spleen)	A	+	+
4	A × B (from spleen)	B	+	+

"+" indicates an immune response; "–" indicates no response

1. In Experiment 1, lymphocytes from both types A and B bone-marrow cells were placed in irradiated type A mice. What happened when either vaccinia-A or vaccinia-B was given to the mice? _____

2. In Experiment 2, both types of bone-marrow lymphocytes were placed in irradiated type B mice. What was the result when either vaccine was administered?

_____

3. In Experiment 3, both types of lymphocytes that had matured in spleen cells were placed in irradiated type A mice. What was the result when either vaccine was administered?

_____

4. In Experiment 4, both types of lymphocytes that matured in spleen cells were donated to irradiated type B mice. Describe their reaction to both kinds of vaccines.

_____

 Copyright © by the Glencoe Division of Macmillan/McGraw-Hill School Publishing Company

Interpreting and Applying Concepts   continued

5. How would you explain the results of these experiments? _____

_____

_____

_____

_____

In other experiments, mice of type C normally reject grafts from type D mice. However, newborn mice of type C can be given cells from type D mice. If six weeks later, type C mice receive a skin graft from mice of type D and E, they accept the grafts from type D mice and reject the grafts from type E mice.

6. Why did the newborn type C mice accept cells from type D mice? _____

_____

7. Why did the six-week old type C mice accept the grafts from type D mice but not the grafts

from type E mice? _____

_____

_____

_____

You have learned that T cells recognize certain self proteins. These proteins are called HLA, which stands for "human leucocyte antigens." Unless you are an identical twin, you have a combination of different HLA on the surface of your cells that is not likely to be duplicated in anyone else. The reason for the variation in HLA is not known, but it has been suggested that by carrying a large number of different HLA, your immune system is less likely to be invaded by a microbe that imitates one of your self proteins. To determine your HLA type, your lymphocytes are examined. Table 23–2 lists some diseases associated with certain HLA. The relative risk column shows the extent to which an individual carrying a particular HLA is more likely to contract the disease when compared to a person without the antigen. Study the table. Then answer the questions that follow.

Table 23–2

Disease	Antigen	Relative Risk
Ankylosing spondylitis	B27	87.8
Rheumatoid arthritis	DRw4	4.0
	DR4	6.0
	B8	1.0
Multiple sclerosis	A3	1.8
	B7	2.0
	Bw2	1.9
	DRw2	3.8

IMMUNE SYSTEM

Interpreting and Applying Concepts   continued

Disease	Antigen	Relative Risk
Addison's disease	Dw3	8.8
	DR3	6.5
	B8	4.0
Grave's disease	B8	2.5
	Bw35	5.0
	Dw3	5.5
Juvenile diabetes	B8	2.7
	DR3	5.0
	DR4	5.0
	DR3 and DR4	14.0

8. How are a person's HLA a defense against disease? _____

_____

9. What disease is associated with HLA B27? What risk is there that a person with B27 will more likely contract the disease than a person who lacks the antigen?

_____

_____

10. How much greater is the risk of developing juvenile diabetes for an individual who has both DR3 and DR4 than for someone who has only one of these antigens?

_____

11. To which diseases is a person with Dw3 susceptible? _____

_____

12. In a population, every individual has his or her own, different HLA combination. How might

this difference in self-proteins help preserve the species? _____

_____

_____

13. If, as self proteins, the HLA help protect the body, hypothesize why the HLA listed in

Table 23–2 are linked with diseases that destroy the body. _____

_____

_____

_____

Copyright © by the Glencoe Division of Macmillan/McGraw-Hill School Publishing Company

CHEMICAL CONTROL

Understanding Concepts

Write the word or phrase that best completes the statement.

1. Any plant response caused by unequal stimulation is called a(n) _____ .

2. _____ are plant growth hormones produced in the tips of stems.

3. _____ occurs when hormones from the tip of a stem inhibit the development of lateral buds.

4. A group of plant hormones called _____ promote dormancy in seeds and buds.

5. In animals, chemical control is regulated by hormones secreted by the _____ .

In the space at the left, write the letter of the phrase from Column B that best matches the term in Column A.

Column A		Column B
_____ 6. gibberellins	**a.**	growth response of a plant to gravity
_____ 7. phototropism	**b.**	hormone that triggers the uptake of glucose from the blood stream
_____ 8. protein-type hormones	**c.**	hormones that cause cell elongation and apical dominance in plants
_____ 9. pituitary gland		
_____ 10. geotropism	**d.**	gland that controls the secretion of hormones from many other endocrine glands
_____ 11. insulin	**e.**	the dropping of fruit or leaves from a plant
_____ 12. abscission	**f.**	cause a response in cells through the action of a second messenger
_____ 13. hypothalamus		
_____ 14. steroid hormones	**g.**	a brain structure that exerts control over the pituitary gland
_____ 15. FSH	**h.**	hormones that activate genes within cells of target tissue
_____ 16. pancreas		
_____ 17. vasopressin	**i.**	growth hormones that cause elongation, increase germination rate and fruit production of some plants
_____ 18. thyroxine	**j.**	growth response of a plant to light
_____ 19. auxins	**k.**	released by the pituitary gland early in the menstrual cycle
	l.	hormone that regulates metabolic rate of body cells
	m.	gland that plays a vital role in regulating blood glucose levels
	n.	hormone that helps maintain the body's water balance

Copyright © by the Glencoe Division of Macmillan/McGraw-Hill School Publishing Company

CHEMICAL CONTROL

Interpreting and Applying Concepts

The drawing in Figure 24–1 represents a plant shoot. Study Figure 24–1 and answer the questions that follow.

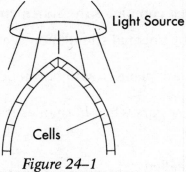

Figure 24–1

1. In the space provided, sketch the same shoot as it would appear after several days of growth. Be sure to include individual cells in your drawing.

2. How would your sketch have been different if the light source had been to the left of the shoot? Directly above the shoot? _____

_____

_____

Answer the following questions in complete sentences.

3. A landscaper wanted to have bushy shrubs in front of a house and conical shrubs along the driveway leading to the house. How could this combination of shapes be achieved using the same kind of plant throughout? _____

_____

_____

4. Why do you think the addition of gibberellins makes many dwarf plants grow tall while it has little effect on plants that are naturally tall or grow as vines?

_____

_____

    Copyright © by the Glencoe Division of Macmillan/McGraw-Hill School Publishing Company

CHEMICAL CONTROL

Interpreting and Applying Concepts   continued

5. What effect might the application of gibberellins to grass seed planted in early spring have on the thickness of the lawn produced from this seed? _____

   _____

6. As part of a negative feedback loop to maintain a constant level of calcium in the blood, the thyroid gland releases a hormone called calcitonin whenever the level of calcium in the blood is raised. Predict what happens next with regard to the calcium level in the blood and the secretion of calcitonin. _____

   _____

7. The pituitary is often referred to as the "master gland." Explain why this is an appropriate name for this gland. _____

   _____

8. A human embryo produces progesterone. What effect does this action have on the menstrual cycle of a pregnant woman? Explain the reason for this effect. _____

   _____

   _____

   _____

9. Hypoglycemia is a condition in which the glucose level in the blood falls to extremely low levels. Describe two ways that this condition might be treated.

   _____

   _____

10. Both human growth hormone (HGH) and thyroxine are needed for normal growth. What are the effects of HGH and thyroxine? How might a lack of thyroxine inhibit a person's growth?

   _____

   _____

   _____

11. In a person with untreated diabetes mellitus, would you expect secretion of glucagon to be increased or decreased? Explain. _____

   _____

   _____

Copyright © by the Glencoe Division of Macmillan/McGraw-Hill School Publishing Company

CHAPTER 25 ❧ ASSESSMENT

NERVOUS CONTROL

Understanding Concepts

In the space at the left, write the letter of the word or phrase that best completes the statement or answers the question.

_____ 1. Which is *not* required in order for a nervous response to occur?
    a. stimulus detection   **c.** hormone secretion
    b. effector response   **d.** impulse transmission

_____ 2. In most animals, incoming impulses are received and transmitted by
    a. sensory neurons.   **c.** interneurons.
    b. motor neurons.   **d.** response neurons.

_____ 3. In a nerve cell, impulses are transmitted away from the cell body by
    a. effectors.   **b.** dendrites.   **c.** myelin.   **d.** an axon.

_____ 4. All of these can affect the speed with which nerve impulses are conducted *except*
    a. myelin sheath.   **b.** nodes.   **c.** polarity.   **d.** axon diameter.

_____ 5. Chemicals involved in moving nerve impulses across a synapse are called
    a. neuroexciters.   **b.** neurotransmitters.   **c.** receptor molecules. **d.** neuroinhibitors.

_____ 6. Chemical substances that reduce nerve transmission across the synapse are called
    a. stimulants.   **b.** depressants.   **c.** neurotoxins.   **d.** inhibitors.

_____ 7. The simplest phylum of animals that possess a nervous system is
    a. monera.   **b.** porifera.   **c.** planaria.   **d.** cnidaria.

_____ 8. The largest part of the human brain is the
    a. medulla.   **b.** cerebrum.   **c.** brainstem.   **d.** cerebellum.

_____ 9. Sympathetic and parasympathetic nerves are controlled by the
    a. medulla oblongata.   **c.** cerebral cortex.
    b. cerebellum.   **d.** thalamus.

_____ 10. The part of the human brain most closely associated with endocrine control is the
    a. cerebral cortex.   **b.** pons.   **c.** hypothalamus.   **d.** midbrain.

_____ 11. Involuntary or automatic responses to stimuli
    a. do not involve the CNS.   **c.** are called reflexes.
    b. are regulated by hormones.   **d.** are controlled by the autonomic system.

_____ 12. Touch is a general term that includes each of these related senses *except*
    a. pain.   **b.** brightness.   **c.** cold.   **d.** pressure.

NERVOUS CONTROL

Interpreting and Applying Concepts

Use the letters of the steps listed below to answer the question that follows. Write the letters in the spaces provided.

    a. impulses cross synapses and activate muscle cell effectors

    b. impulses cross synapses and enter spinal cord interneurons

    c. muscles contract, moving the hand

    d. impulses interpreted as heat and pain

    e. some impulses cross synapses to motor neurons, which leave the spinal cord

    f. sensory neurons are stimulated

    g. receptors in the skin are activated

    h. some impulses travel to the brain

    i. impulses travel to the spinal cord

1.   If your hand touches a hot stove, the reflex response occurs in two stages—initial and secondary. In what order do the steps occur?

    initial: ___ ⟶ ___ ⟶ ___ ⟶ ___ ⟶ ___ ⟶ ___ ⟶ ___

                 secondary: ⟶ ___ ⟶ ___

Answer the following questions in complete sentences.

2.   The human tongue can detect four different taste sensations—sweet, salt, acid, and bitter. These sensations are detected by sensors called taste buds. A particular type of taste bud is specific for one taste sensation only. For example, taste buds that are stimulated by sweet substances are not stimulated by sour, salty, or bitter substances. Different types of taste buds are concentrated in different regions of the tongue.

        Describe an experiment that could help you "map" the locations of these different taste regions on the tongue.

_____

_____

_____

_____

Copyright © by the Glencoe Division of Macmillan/McGraw-Hill School Publishing Company

NERVOUS CONTROL

Interpreting and Applying Concepts   continued

3. Why is it important that reflex actions be involuntary rather than voluntary?

_____

_____

_____

_____

4. Predict the initial and secondary reflex responses to hitting your right thumb with a hammer.

_____

_____

5. Acetylcholine causes muscles to contract and also to relax. Explain this seeming contradiction.

_____

_____

_____

_____

_____

_____

_____

6. Some poisons, such as curare, bind to the receptors for the neurotransmitter acetylcholine on muscles, inhibiting these muscles. What effect does this action have on the human body?

_____

_____

_____

7. People who have lost an arm or leg due to accident or surgical amputation often report feeling sensations at the site of the missing body part. What is a likely explanation for this

phenomenon? _____

_____

_____

Copyright © by the Glencoe Division of Macmillan/McGraw-Hill School Publishing Company

CHAPTER 26 🐾 ASSESSMENT

MOVEMENT

Understanding Concepts

Answer the following questions.

1. What material makes up most of the skeleton of the fetus of a vertebrate? _____

2. What structural characteristic is unique to vertebrates? _____

3. Why do arthropods molt? _____

4. Name three important functions of an exoskeleton. _____

5. What are the individual parts that make up an endoskeleton? _____

6. What is the process by which cartilage is replaced by bone? _____

7. What are two nutrients needed for proper bone development? _____

8. Where are Haversian canals found and what is their function? _____

   _____

9. What are the connective tissues of a moveable joint? _____

10. Name the four types of moveable joints in vertebrates. _____

    _____

In the space at the left, write TRUE if the statement is true. If the statement is false, change the italicized word or phrase to make it true.

_____ 11. In a human skeleton, fixed joints are found in the *wrist and ankle*.

_____ 12. When a person's triceps muscle contracts, his or her arm *straightens*.

_____ 13. Muscles that move the skeleton are made up of *cardiac* muscle tissue.

_____ 14. In addition to vertebrates, *cnidarians* make up the only other animal group to produce movement through the combined actions of muscles and skeletons.

_____ 15. The sliding filament hypothesis explains the process of *muscle contraction*.

_____ 16. Nerves of the autonomic system control the action of *voluntary* muscles.

_____ 17. The human hip is an example of a *ball-and-socket* joint.

_____ 18. Bones are formed by special cells called *osteoblasts*.

MOVEMENT

Interpreting and Applying Concepts

Figure 26–1 shows a working model constructed
by a student to show how opposing muscles
produce motion around a joint. The model
consists of two pieces of wood joined together
with elastic tape, a colored balloon, a white balloon,
and some velcro strips used to attach the balloons
to the pieces of wood. Study Figure 26–1
and answer the questions that follow.

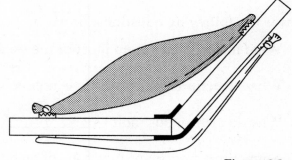

Figure 26–1

1. What do the pieces of wood represent? _____

2. What kind of joint is represented? _____

3. What does the elastic tape represent? _____

4. What do the velcro strips represent? _____

5. What kind of muscle is represented by the colored balloon? The white balloon?

   _____

6. Where are similar joints and muscle pairs found in the human body? Name the joints.

   _____

7. What effect would detaching one of the velcro strips from the white balloon have on the ability

   of the muscle pair to perform their tasks? _____

   _____

8. Suppose you were asked to design an experiment to test the hypothesis that muscle fatigue

   affects coordination. Describe your experiment. _____

   _____

   _____

   _____

   _____

   _____

Copyright © by the Glencoe Division of Macmillan/McGraw-Hill School Publishing Company

MOVEMENT

Interpreting and Applying Concepts   continued

Answer the following questions in complete sentences.

9. Although the skeletal system consists largely of hard, bony tissue, it also contains a considerable amount of softer, more flexible tissues. Name two types of soft tissue found in the skeletal system and describe the function of each.

   _____

   _____

   _____

10. Most vertebrate organisms are weakest and most vulnerable to injury or capture by predators when they are very young. They become less vulnerable as they grow larger and stronger. Arthropods, on the other hand, remain vulnerable at times well into adulthood. Explain.

    _____

    _____

    _____

11. Describe some methods of movement used by organisms that may lack skeletons or muscles

    or both. _____

    _____

    _____

    _____

12. In humans, the Achilles tendon, located at the lower back portion of the leg, attaches the calf muscles to the heel bone of the foot. Because of its location, the Achilles tendon is vulnerable to injury, even to being severed, especially during athletic events. How would severing the

    Achilles tendon affect body movement? _____

    _____

    _____

    _____

13. How might the movement of the human body be affected if striated muscle tissues were replaced with smooth muscle tissues over which one had voluntary control?

    _____

    _____

POPULATION BIOLOGY

Understanding Concepts

In the space at the left, write the letter of the phrase from Column B that best matches the term in Column A.

Column A

_____ 1. social hierarchy

_____ 2. biotic potential

_____ 3. population density

_____ 4. interspecific competition

_____ 5. intraspecific competition

_____ 6. limiting factors

_____ 7. carrying capacity

_____ 8. population growth curve

_____ 9. emigration

_____ 10. predation

Column B

a. circumstances that prevent organisms from reaching their biotic potential

b. a designated chain of command based on dominance

c. the moving of a population out of an area

d. the highest rate of reproduction under ideal conditions

e. a population graph having an S shape

f. competition among populations of different species

g. the feeding of one organism on another

h. the maximum number of individuals in a given population that the environment can support

i. the size of a population that occupies a given area at any given point in time

j. competition between members of the same species

Write the word or phrase that best completes the statement.

11. _____ is a condition in which the birthrate equals the death rate and the population is not growing in size.

12. Social hierarchy is also known as a _____ .

13. _____ occurs when a species dies out due to competition with a different species for the same resources.

14. Food supply and living space are examples of _____ limiting factors.

15. The swarming of bees from an overcrowded hive to a new location is an example of

_____ .

Copyright © by the Glencoe Division of Macmillan/McGraw-Hill School Publishing Company

POPULATION BIOLOGY

Interpreting and Applying Concepts

Each of the graphs in Figure 27–1 shows a relationship between change in population size and time. Refer to the graphs as you answer the following questions.

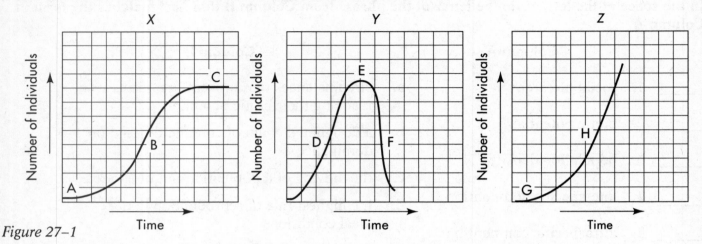

Figure 27–1

1. Which graph represents growth under ideal conditions? How do you know?

   _____

2. Describe what is happening in graph Y to the population size at point F.

   _____

3. In graph X, what is happening to the population size at points A, B, and C?

   _____

4. How does the birthrate of the organisms compare with the death rate at point C? At point F?

   _____

5. If each graph were extended beyond the time limits shown, which would show a continued growth in population size?

   _____

6. The curve of which graph most closely resembles that expected for a frog population in a pond? For a lightning bug population?

   _____

   _____

7. Which graph shows the effects of density-dependent limiting factors on population size? What might some of those limiting factors be?

   _____

   _____

          Copyright © by the Glencoe Division of Macmillan/McGraw-Hill School Publishing Company

POPULATION BIOLOGY

Interpreting and Applying Concepts   continued

Answer the following questions in complete sentences.

8. Both interspecific competition and intraspecific competition are density-dependent limiting factors. Why is extinction a possible outcome of interspecific competition, but not of intraspecific competition?

_____

_____

_____

_____

_____

Questions 9 and 10 are related questions. Where appropriate, use the terms *limiting factors,* *carrying capacity,* **and** *population density* **in your answer.**

9. Suppose you were to place several mating pairs of rabbits on a small island having ample vegetation and no predators. Describe the changes you would expect to see in the rabbit population over time. Include a discussion of any limiting factors in your description.

_____

_____

_____

_____

_____

10. Once the island's carrying capacity for rabbits is reached, several mating pairs of foxes are introduced to the island. What effect will this have on the rabbit population of the island? Describe the changes in rabbit and fox populations you would expect to take place over time.

_____

_____

_____

_____

_____

_____

POPULATION BIOLOGY

Interpreting and Applying Concepts   continued

The age structure diagrams in Figure 27–2 show how individuals are distributed at each age level for different human populations. Study the diagrams and then answer the questions.

Figure 27–2

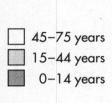

45–75 years
15–44 years
0–14 years

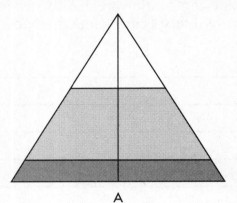

A

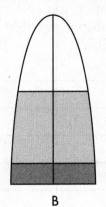

B

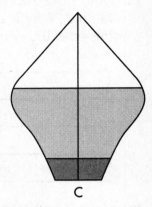

C

11. Based on what you know about average reproductive age for humans, which graph do you think represents a rapidly expanding population? Explain.

_____

_____

_____

12. Describe the types of populations represented by the two graphs you did not name in your answer to Question 11.

_____

   Copyright © by the Glencoe Division of Macmillan/McGraw-Hill School Publishing Company

CHAPTER 28 ❧ ASSESSMENT

ECOSYSTEMS

Understanding Concepts

In the space at the left, write the letter of the word or phrase that best completes the statement or answers the question.

_____ 1. All the possible feeding levels within an ecosystem make up
    a. a food chain.   **b.** a biome.   **c.** abiotic factors.   **d.** a food web.

_____ 2. A feeding step within an ecosystem is also known as
    a. a pyramid.   **b.** a trophic level.   **c.** a niche.   **d.** a producer.

_____ 3. Relationships between or among different organisms in an environment are known as
    a. interactive biomass.   **c.** diversifications.
    b. biotic factors.   **d.** communal activities.

_____ 4. A rabbit that eats a plant is
    a. a first-order consumer.   **c.** a second-order consumer.
    b. a producer.   **d.** a carnivore.

_____ 5. All first-order consumers are
    a. carnivores.   **b.** omnivores.   **c.** small animals.   **d.** herbivores.

_____ 6. The pyramid of numbers does not apply to food chains in which
    a. consumers are at the lowest trophic level.
    b. carnivores outnumber herbivores.
    c. producers are at the higher trophic levels.
    d. small organisms feed on a large organism.

_____ 7. What type of relationship between two organisms results in benefits to both organisms?
    a. parasitism   **b.** mutualism   **c.** equalism   **d.** commensalism

_____ 8. All of the interacting biotic and abiotic factors in an environment make up
    a. a food chain.   **b.** a population.   **c.** an ecosystem.   **d.** a biomass pyramid.

Answer the following questions.

9. What term is used to identify an organism's place in an ecosystem? _____

10. What happens to most of the potential energy that is "lost" between links of a food chain?

_____

11. What is the term used to identify organisms that break down and consume organic materials

and wastes in a food chain? _____

12. What is the trophic level of a person eating a piece of bread? _____

ECOSYSTEMS

Interpreting and Applying Concepts

Answer the following questions.

1. In the space provided, construct a food web made up of the following organisms: hawk, field mouse, deer, grass, shrubs, cricket, rabbit, frog, mountain lion. Then describe the food chains in your food web.

_____

_____

_____

_____

_____

2. Consider a hypothetical situation in which 20 000 kilocalories of energy is available to the producers in a food chain. First-order consumers receive 20 percent of this energy, and the available energy is reduced by half for each of the next two levels. Determine the amount of energy available at the first, second, and third trophic levels of this food chain.

_____

_____

_____

Answer the following questions in complete sentences.

3. Many conservation-minded people use laundry detergents that do not contain phosphates. How does such a practice help the environment?

_____

_____

_____

_____

_____

_____

_____

 Copyright © by the Glencoe Division of Macmillan/McGraw-Hill School Publishing Company

ECOSYSTEMS

Interpreting and Applying Concepts   continued

4. Compare and contrast commensalism, mutualism, and parasitism.

_____

_____

_____

_____

_____

_____

_____

5. Why are fuels such as coal, natural gas, and petroleum products called fossil fuels? Trace the energy stored in these fuels back to its original source—energy from the sun.

_____

_____

_____

_____

_____

_____

6. How would a food web in a desert area be similar to one in a tropical rain forest? How would it be different? _____

_____

_____

7. Only a small percentage of the sun's energy that reaches Earth's surface is converted to chemical energy through photosynthesis. What happens to most of the sun's energy that reaches Earth and how does this affect Earth's environments?

_____

_____

8. Why is the diversity of life greater near the shorelines and near the water's surface than in deeper waters? _____

_____

_____

_____

Copyright © by the Glencoe Division of Macmillan/McGraw-Hill School Publishing Company

ECOSYSTEMS

Interpreting and Applying Concepts   continued

9. Loam is a soil consisting of a mixture of clay, silt, sand, and a small amount of gravel. A topsoil consisting of loam mixed with humus is more fertile than topsoils consisting of humus mixed with clay or with sand alone. Explain.

_____

_____

_____

_____

10. Instead of leaving land bare, farmers often plant legumes, such as clover or alfalfa, on land that is not being used to grow cash crops. Why?

_____

_____

_____

_____

  Copyright © by the Glencoe Division of Macmillan/McGraw-Hill School Publishing Company

Name _____ Date _____ Class _____

ORIGIN AND DISTRIBUTION OF COMMUNITIES

Understanding Concepts

In the space at the left, write the letter of the biome from the following list that best matches each description. A biome may be used more than once.

a. desert   **b. grassland**   **c. taiga**   **d. temperate forest**   **e. tropical rain forest**   **f. tundra**

_____ 1. coniferous trees

_____ 2. permanently frozen soil

_____ 3. prairie, steppe, pampa, veldt

_____ 4. little rainfall; rapid evaporation

_____ 5. deciduous trees

_____ 6. low temperature; short summer

_____ 7. uneven rainfall; scattered trees

_____ 8. fog; water-soaked, acidic soil

_____ 9. woodland with definite seasons

_____ 10. constant temperature; heavy rainfall

In the space at the left, write the letter of the word or phrase that best completes the statement or answers the question.

_____ 11. Two abiotic factors affecting the distribution of organisms in the ocean are
    a. light and temperature.
    b. humidity and altitude.
    c. temperature and precipitation.
    d. light and latitude.

_____ 12. Each of the following is characteristic of a climax community *except*
    a. large plants with long life cycles.
    b. steady population sizes.
    c. simple food chains.
    d. diversity of species.

_____ 13. The top layer of vegetation in a temperate or tropical rain forest is the
    a. biosphere.   **b.** benthos.   **c.** canopy.   **d.** shrub layer.

_____ 14. The circulation of water in the ocean is mostly affected by water temperature and the
    a. movement of ocean animals.
    b. tilt of Earth's axis.
    c. topography.
    d. wind.

_____ 15. A possible sequence of plants in primary land succession is
    a. grasses , shrubs, mosses, trees, ferns.
    b. lichens, mosses, grasses, shrubs, trees.
    c. trees, mosses, lichens, shrubs, grasses.
    d. shrubs, grasses, ferns, mosses, lichens.

In the space at the left, write **TRUE** if the statement is true. If the statement is false, change the italicized word or phrase to make it true.

_____ 16. The *biosphere* is a layer of soil that never thaws during the tundra summer.

_____ 17. Desert communities tend to form on the *leeward* side of mountains.

_____ 18. *Plankton* move freely through the ocean under their own power.

Copyright © by the Glencoe Division of Macmillan/McGraw-Hill School Publishing Company

ORIGIN AND DISTRIBUTION OF COMMUNITIES

Interpreting and Applying Concepts

A climatogram is a graph that displays average monthly temperature and precipitation for a particular region. The months of the year are given along the bottom. The bar graph indicates average monthly precipitation and uses the scale along the left side. The line graph indicates average monthly temperature and uses the scale along the right side.

The table below gives the average annual precipitation and temperature range for the five biomes you studied. Use the table and climatograms in Figure 29-1 to answer the questions.

Biome	Average Annual Precipitation (cm)	Range of Average Monthly Temperatures (°C)
Tundra	less than 25	–24° to 4°
Temperate Forest	75 to 100	5° to 29°
Tropical Rain Forest	more than 200	24° to 28°
Grassland	25 to 75	0° to 24°
Desert	less than 25	25° to 33°

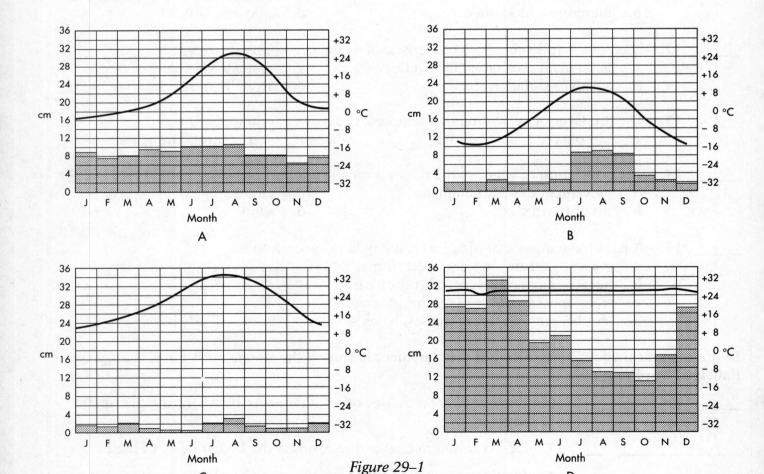

Figure 29–1

   Copyright © by the Glencoe Division of Macmillan/McGraw-Hill School Publishing Company

ORIGIN AND DISTRIBUTION OF COMMUNITIES

Interpreting and Applying Concepts   continued

1. What is the range of the average monthly temperatures of the region shown in climatogram A?

   _____

2. What is the total annual precipitation of the region shown in climatogram C?

   _____

3. Based on precipitation, what two biomes could climatogram C represent?

   _____

4. What other factor(s) could you use to determine which biome climatogram C represents?

   _____

5. What type of biome does climatogram B represent? Explain.

   _____

   _____

6. Name some animals that are likely to inhabit the region shown in climatogram B.

   _____

7. Which climatogram represents a tropical rain forest biome? Explain your reasoning.

   _____

   _____

8. How might a climatogram representing a grassland biome compare with climatogram A?

   _____

   _____

   _____

ORIGIN AND DISTRIBUTION OF COMMUNITIES

Interpreting and Applying Concepts continued

The hot, sandy beach area along a lake or ocean shore may be marked by sand dunes built up by the action of the wind. The four illustrations in Figure 29–2 show the primary succession in such a sand dune community, but the stages are not shown in order. Study Figure 29-2, then answer the questions.

Figure 29–2

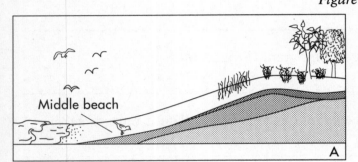
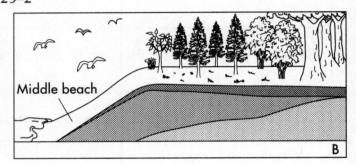
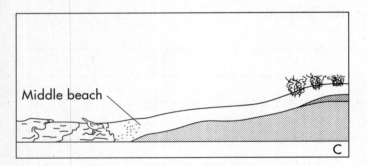
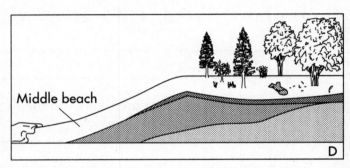

☐ Beach Sand   ▨ Sand Added By Waves and Wind   ▧ Sandy Topsoil

9. Sequence the stages of succession from pioneer stage to climax community. _____

10. What factors helped you identify the pioneer stage? _____

11. How did you identify the climax stage? _____

12. What types of animals are likely to inhabit the area during the stage shown in illustration A?

_____

13. What factors might explain why no vegetation develops in the middle beach area throughout the stages of succession?

_____ _____

14. How do the pioneer plants in the sand dune succession differ from the pioneer plants in the forest succession shown on page 816 of your textbook? _____

_____

Copyright © by the Glencoe Division of Macmillan/McGraw-Hill School Publishing Company

CHAPTER 30 ASSESSMENT
HUMANS AND THE ENVIRONMENT

Understanding Concepts

In the space at the left, write TRUE if the statement is true. If the statement is false, change the italicized word or phrase to make it true.

_____ 1. *Habitat destruction* is a major cause of the extinction of plant and animal species.

_____ 2. The removal of soil by the action of moving air and water is *pollution*.

_____ 3. Compounds that cannot be broken down into inactive chemicals by living organisms are *nonrenewable*.

_____ 4. *Chlorofluorocarbons* are chemicals used to kill unwanted organisms.

_____ 5. Soil that is no longer fertile enough to support crop growth because its nutrient content has been reduced is said to be *eroded*.

_____ 6. Integrated pest management includes the breeding of pest-resistant plants and the *use of pesticides*.

In the space at the left, write the letter of the possible solution or alternative from the following list that matches the given environmental problem. A problem may have more than one solution, and a solution may be used more than once.

a. alternative energy sources   b. autocidal controls   c. ban CFC use   d. biological controls
e. coal scrubbers   f. contour plowing   g. crop rotation   h. reforestation   i. terracing

_____ 7. acid precipitation   _____ 11. pesticide use

_____ 8. depletion of ozone   _____ 12. photochemical smog

_____ 9. erosion   _____ 13. soil depletion

_____ 10. greenhouse effect

In the space at the left, write the letter of the word or phrase that best completes the statement or answers the question.

_____ 14. Each of the following is an alternative to pesticide use *except*
   a. autocidal control.   c. biological control.
   b. biological magnification.   d. cultural control.

_____ 15. Food and forests are considered
   a. nonrenewable resources.   c. biodegradable.
   b. nonbiodegradable.   d. renewable resources.

_____ 16. The alternate planting of soil-enriching and soil-depleting crops on an area of land is
   a. crop rotation.   c. terracing.
   b. strip cropping.   d. the greenhouse effect.

HUMANS AND THE ENVIRONMENT

Interpreting and Applying Concepts

Pesticide X is a nonbiodegradable pesticide that was used to treat plants in a forest community. Study the chart and food web shown in Figure 30–1, then answer the questions.

Representative Concentrations of Pesticide X in Tissues (parts per million, ppm)	
Producers	0.01–0.05
First-order consumers	0.25–1.5
Second-order consumers	2.1–4.5
Third-order consumers	4.1–13.8
Fourth-order and above	4.5 +

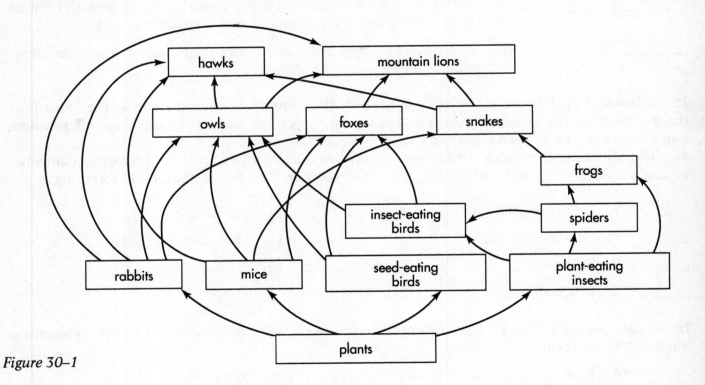

Figure 30–1

1. About how many times higher is the average concentration of pesticide X in primary consumers than in producers? _____

2. Which animal(s) in the food web will have the highest concentrations of pesticide X in their bodies? _____

  Copyright © by the Glencoe Division of Macmillan/McGraw-Hill School Publishing Company

HUMANS AND THE ENVIRONMENT

Interpreting and Applying Concepts   continued

3. Would a snake that primarily feeds on frogs tend to have a higher or lower concentration of pesticide X in its tissues than a fox that primarily feeds on rabbits? Explain your reasoning.

_____

_____

_____

4. If humans ate rabbits, snakes, and frogs, about how much pesticide X (in parts per million) would you expect to find in the tissues of a human that was part of this food web?

_____

Complete Table 30–1 and answer the questions.

Table 30–1

Pollutant	Sources	Negative Effect on Environment
carbon monoxide		
carbon dioxide		
sulfur dioxide		
nitrogen oxides		
CFCs		
pesticides		
hydrocarbons		
biological wastes		

5. What is the major contributor to the pollution of the environment?

_____

Copyright © by the Glencoe Division of Macmillan/McGraw-Hill School Publishing Company

HUMANS AND THE ENVIRONMENT

Interpreting and Applying Concepts   continued

6. What part of the environment (air, water, soil) is most affected by carbon dioxide? What part is most affected by pesticides? _____

7. How do sulfur dioxide and nitrogen oxides pollute our waters? _____

_____

_____

_____

8. Fertilizers, which are used to add nutrients to depleted soil, can become pollutants. What part(s) of the environment would be affected by fertilizer use? _____

9. If fertilizers run off into lakes, how could they affect a lake ecosystem? _____

_____

_____

_____

_____

10. Explain why ozone formation and ozone depletion both have negative effects on the environment.

_____

_____

_____

    Copyright © by the Glencoe Division of Macmillan/McGraw-Hill School Publishing Company

CHAPTER 1 ✎ ASSESSMENT

BIOLOGY—THE SCIENCE OF LIFE

Understanding Concepts

Classify each organism as a producer (p), consumer (c), or decomposer (d) by writing the correct letter in the space at the left.

p 1. algae   **c** 5. garden snake   **c** 9. tiger

c 2. ant   **c** 6. goldfish   **p** 10. tree

d 3. bread mold   **p** 7. grass

c 4. cow   **d** 8. mushroom

In the space at the left, write the letter of the word or phrase that best completes the statement or answers the question.

c 11. Two products of cellular respiration are
    a. oxygen and glucose.   c. carbon dioxide and water.
    b. oxygen and water.   d. carbon dioxide and glucose.

b 12. Each of the following is a biologically-based solution to controlling the zebra mussel invasion *except*
    a. depriving them of oxygen.   c. introducing one of their predators.
    b. scraping them off surfaces.   d. preventing fertilization of their eggs.

b 13. In producers, chlorophyll and sunlight are necessary for the process of
    a. homeostasis.   c. cellular respiration.
    b. photosynthesis.   d. reproduction.

d 14. The closing of its shell when a clam is removed from its watery environment is an example of how the clam maintains its
    a. growth.   b. development.   c. evolution.   d. homeostasis.

a 15. In a food chain involving a mouse and the snake that eats it, the mouse and snake are
    a. both consumers.   c. consumer and decomposer, respectively.
    b. both producers.   d. producer and consumer, respectively.

Write the word or phrase that best completes the statement.

16. The supply of energy that producers need to make food comes from ___ **the sun** ___ .

17. Physical traits that make an organism well suited to its environment are called ___ **adaptations** ___ .

18. The broadest division into which organisms may be classified is a(n) ___ **kingdom** ___ .

19. The increase in the amount of living material in an organism is referred to as ___ **growth** ___ .

20. The changes an organism undergoes in reaching its adult form is its ___ **development** ___ .

Copyright © by the Glencoe Division of Macmillan/McGraw-Hill School Publishing Company

---

BIOLOGY—THE SCIENCE OF LIFE

Interpreting and Applying Concepts

The graphs in Figure 1–1 depict human growth in height from birth through age 18 for males and females. Study the graphs and then answer the questions.

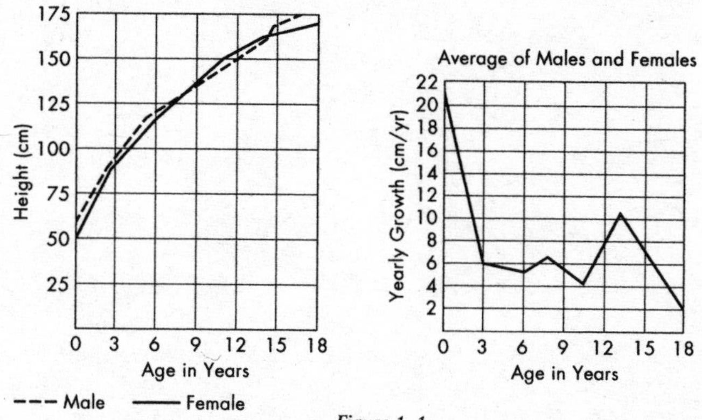

Figure 1–1

1. On the first graph, what indicates that growth takes place from birth to age 3? **The lines slope upward and the values increase from about 50 cm at birth to about 95 cm at age 3 for both males and females.**

2. From ages 6 to 9, about how many centimeters does a female child grow? **about 15 cm**

3. During what range of ages do females generally grow faster than males? **from around age 9 to age 14**

4. About how many centimeters taller are males at age 15 than females at the same age? **about 5 cm**

5. During which years do all children tend to grow the fastest? **from birth to age 3**

6. Describe the trend in human growth depicted in the second graph. **Humans show the most growth during the first 3 years. Growth seems to slow down from ages 3 through 11 or 12. At around ages 12-15 another period of rapid growth takes place.**

Copyright © by the Glencoe Division of Macmillan/McGraw-Hill School Publishing Company

121

BIOLOGY–THE SCIENCE OF LIFE

Interpreting and Applying Concepts continued

One food chain based on this model consists of corn grain, a corn-fed chicken, and a human who eats chickens. Suppose the corn contains 100 energy units.

10. How many energy units are available to the chicken from corn? __10 units__

11. Which is a more energy-efficient diet for humans, a diet of corn or a diet of chicken? Explain. **A diet of corn is more energy-efficient because chickens are fed corn, and a lot of the energy in the corn is lost as heat. Therefore, a lot of corn must first be raised to feed the chickens.**

12. Which would support a larger number of people with less energy input, a diet of corn or a diet of corn-fed chicken? Explain your reasoning. **A diet of corn because it takes more energy to produce chicken than to produce corn.**

Copyright © by the Glencoe Division of Macmillan/McGraw-Hill School Publishing Company

---

BIOLOGY–THE SCIENCE OF LIFE

Interpreting and Applying Concepts continued

7. Complete Table 1-1 comparing cellular respiration and photosynthesis.

Table 1-1

Cellular Respiration	Photosynthesis
energy is given off	**energy is stored**
occurs in plants and animals	occurs in green plants
produces carbon dioxide and water	**produces oxygen and simple sugar**
simple sugar and oxygen combine	carbon dioxide and water combine

Figure 1-2 shows the number of usable energy units available at each level of a food chain. Use the figure to answer the questions.

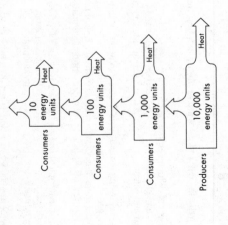

Consumers — 10 energy units — Heat
Consumers — 100 energy units — Heat
Consumers — 1,000 energy units — Heat
Producers — 10,000 energy units — Heat

Figure 1-2

8. How many energy units are lost in the form of heat as energy flows from producer to first consumer in this food chain? What percent of the original amount does that represent? **9000 units; 90%**

9. As you go up the food chain, how does the amount of energy available at each level compare with the previous level? **At each successive level, there is 1/10 as much energy as at the previous level.**

Copyright © by the Glencoe Division of Macmillan/McGraw-Hill School Publishing Company

122

BIOLOGY AS A SCIENCE

Interpreting and Applying Concepts

Based on her field observations, a scientist hypothesized that a certain species of insect requires a specific range of air temperatures in order to live. She tested her hypothesis by exposing the larval stage of this species to varying air temperatures. The data from her experiments is shown below. Use the data to answer the following questions.

TEMPERATURE (°C)	SURVIVAL RATE (%)
15	0
16	20
17	60
18	80
19	90
20	100
21	100
22	80
23	70
24	30
25	0

1. What is the independent variable in the experiment? **air temperature**

2. What is the dependent variable in the experiment? **survival rate**

3. What type of graph would you use to display these data? Give reasons for your choice.
Answers may vary, but many students will select a line graph because it can be used to show how the survival rate varies according to temperature.

4. Draw the graph you described in Question 3. **Graphs may vary.**

Insect Survival Rate

Survival Rate (%)

Temperature (°C)

Copyright © by the Glencoe Division of Macmillan/McGraw-Hill School Publishing Company

---

CHAPTER 2 ❧ ASSESSMENT
BIOLOGY AS A SCIENCE

Understanding Concepts

In the space at the left, write the letter of the phrase from Column B that best matches the term in Column A.

	Column A		Column B
d	1. control group	a.	applied science
b	2. data	b.	information gathered from observations
e	3. experiment	c.	process that produces a body of knowledge about nature
h	4. experimental group	d.	group in which all the variables remain constant
f	5. hypothesis	e.	tests an explanation
c	6. science	f.	statement that explains and relates data
a	7. technology	g.	hypothesis that has withstood the test of time
g	8. theory	h.	group in which the independent variable is changed

In the space at the left, write the letter of the word or phrase that best completes the statement or answers the question.

c 9. A good way of showing comparisons among groups when data are not dependent on one another is a
a. line graph.   b. pie graph.   c. bar graph.

b 10. A three-dimensional image of an object is produced by a
a. transmission electron microscope.
b. scanning electron microscope.
c. compound light microscope.

a 11. The length times the width times the height of a bricklike cell is the cell's
a. solid volume.   b. weight.   c. area.

b 12. A mass of 726 grams is equal to
a. 7.26 kilograms.   b. 0.726 kilogram.   c. 72.6 kilograms.

c 13. The starting point of all scientific research is a(n)
a. controlled experiment.   b. hypothesis.   c. observation.

BIOLOGY AS A SCIENCE

Interpreting and Applying Concepts   continued

Use the graph constructed for Question 4 to answer the following questions.

5. What relationship does the graph illustrate? **The graph illustrates the relationship that exists between air temperature and survival rate of a particular species of insect.**

6. At what air temperature(s) did all the organisms survive? **20°C, 21°C**

7. At what air temperature(s) did all the organisms die? **15°C, 25°C**

8. Do the data support the scientist's original hypothesis? Explain. **Yes, the data show that this species of insect does require a specific range of air temperatures in order to survive.**

9. Based on the data, what range of air temperatures offers these organisms the best chance for survival? **Air temperatures that range from 18°C to 22°C.**

10. Based on the data, what range of air temperatures seems to pose the greatest danger to these organisms? **Air temperatures below 18°C or higher than 22°C.**

11. What would you expect to happen to a group of these organisms if the air temperature was 28°C? **All the organisms would probably die.**

12. In conducting her research, the scientist used heat lamps to control the air temperature surrounding the organisms. No variable other than temperature of air was noted. Can she be sure that the results are linked to air temperature? **No; factors such as humidity and radiation from heat lamps might have affected the results.**

13. Suppose you wanted to repeat this experiment to verify the results. Explain how you would go about investigating the original hypothesis. **Answers will vary but should include creating a control group to negate the effect of other variables.**

Copyright © by the Glencoe Division of Macmillan/McGraw–Hill School Publishing Company

CHAPTER 3 ASSESSMENT
MATTER AND ENERGY

Understanding Concepts

In the space at the left, write **TRUE** if the statement is true. If the statement is false, change the italicized word or phrase to make it true.

an acid   1. A solution in which the concentration of hydrogen ions is greater than the concentration of hydroxide ions is *a base*.

A covalent bond   2. *An ionic bond* results when atoms combine by sharing electrons.

true   3. A *chemical formula* shows the number and kind of each atom in a compound.

nucleic acids   4. DNA and RNA belong to the class of biological compounds called *lipids*.

The terms in the first pair are related to each other. In the space at the left, write the letter of the second pair of terms that are related in the same way.

First Pair

                Second Pair

b   5. atom, element
- a. electron, nucleus
- b. molecule, compound
- c. proton, energy level

c   6. glucose, carbohydrate
- a. hemoglobin, nucleic acid
- b. maltose, protein
- c. cholesterol, lipid

a   7. ammonia, base
- a. pure water, neutral
- b. coffee, neutral
- c. blood, acid

b   8. structural formula, organic compound
- a. carbon; hydrogen
- b. plans, house
- c. isomer, organic molecule

In the space at the left, write the letter of the word or phrase that best completes the statement or answers the question.

a   9. The first energy level of an atom holds a maximum of
- a. two electrons.
- b. eight electrons.
- c. 10 electrons.
- d. 18 electrons.

d   10. Atoms of a particular element containing different numbers of neutrons are called
- a. elements.
- b. ions.
- c. crystals.
- d. isotopes.

b   11. The formula for a molecule of carbon dioxide, CO_2, indicates a total of
- a. two atoms.
- b. three atoms.
- c. four atoms.
- d. six atoms.

MATTER AND ENERGY

Interpreting and Applying Concepts

Answer the following questions.

1. A Venn diagram is a visual model used to illustrate the similarities and differences that exist among different groups. Compare the properties of biologically important organic compounds listed below by completing the Venn diagram (Figure 3–1). Fill in the nonoverlapping portion of each circle with the letters of the unique characteristics of carbohydrates, lipids, and proteins that are given in the list. Fill in the space where all the circles overlap with the letters of those characteristics that are common to all types of biologically important organic compounds. One of the letters for the overlapping region has been given.

Figure 3–1

a. not normally used as a source of energy
b. contain nitrogen and sulfur
c. have a ratio of two hydrogen atoms to each oxygen atom
d. often used as long-term energy reserves
e. not soluble in water
f. are organic compounds
g. number of hydrogen atoms per molecule is much greater than the number of oxygen atoms
h. often serve as immediate energy sources for life processes
i. built from amino acids
j. contain carbon, hydrogen, and oxygen atoms

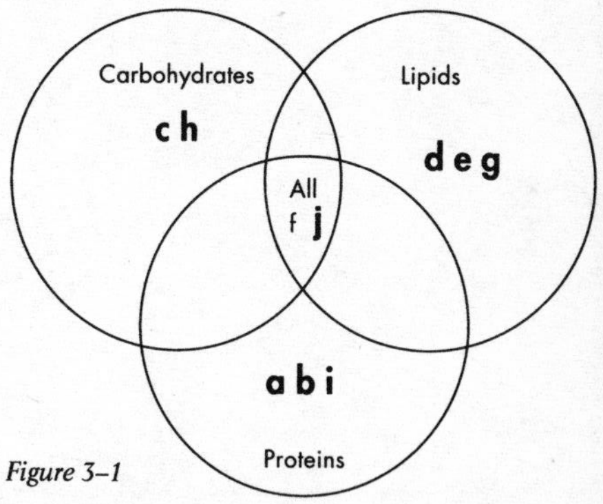

2. To which group of biologically important organic compounds do glycogen and cellulose belong? **carbohydrates**

3. Compare the structure of glycogen with that of cellulose. **Glycogen and cellulose are alike in that both are made of chains of glucose molecules. They differ in the patterns by which the glucose molecules are linked.**

4. The glucose molecules that make up glycogen are joined by alpha bonds, while cellulose contains beta bonds. Cellulose cannot be digested by enzymes in the human body. If both of these compounds are chains of similar molecules, why can't both be digested by the body? **The enzymes in the body's cells are specific--each guides only one type of cell reaction. Evidently, there is a specific enzyme that fits with glycogen and breaks down this substrate. Cellulose has no enzyme that is specific for it, and, because of its different bonding, will not act as a substrate for the enzyme that breaks down glycogen.**

125

MATTER AND ENERGY

Interpreting and Applying Concepts continued

Refer to Figure 3–2 to answer the questions that follow.

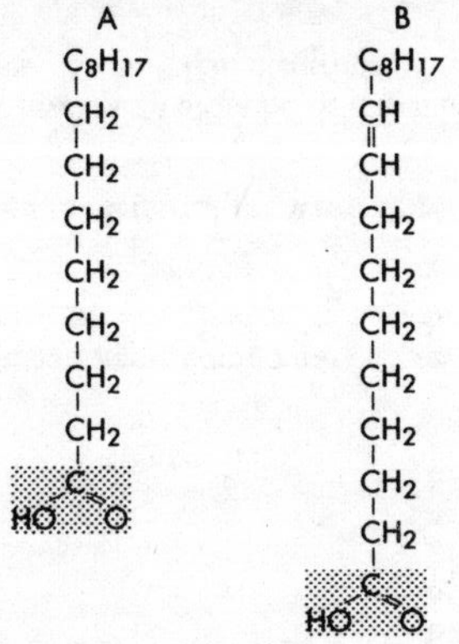

Figure 3–2

5. What type of biological compounds are illustrated in Figure 3–2? __fatty acids__

6. How are the structures of the compounds in Figure 3–2 similar? **Both compounds contain a carboxyl group (–COOH) and chains of linked carbon atoms.**

7. How are the structures of the compounds in Figure 3–2 different? **They differ in that one contains a double bond between 2 carbon atoms.**

8. Classify each compound in Figure 3–2 as either saturated or unsaturated. Provide evidence for your classification. **Compound A is saturated while Compound B is unsaturated. This is due to the fact that Compound A contains single covalent bonds while Compound B contains a double covalent bond.**

9. Do both compounds illustrated in Figure 3–2 contain the maximum number of hydrogen atoms possible? Explain your answer. **No. Compound B contains one double covalent bond and therefore could hold 2 additional hydrogen atoms.**

10. Certain fats are said to be "polyunsaturated." What would you find upon examining the structural formula of such a compound? **It would contain a polyunsaturated fatty acid, which contains 2 or more double covalent bonds in its carbon chain.**

CHAPTER 4 ❧ ASSESSMENT

THE CELL AND ITS ENVIRONMENT

Understanding Concepts

In the space at the left, write **TRUE** if the statement is true. If the statement is false, change the italicized word or phrase to make it true.

exocytosis 1. Cell products or wastes are released to the surroundings through the process of *endocytosis*.

true 2. The *plasma membrane* of a cell regulates which particles enter and leave the cell.

true 3. The diffusion of water into and out of cells across a selectively permeable membrane is called *osmosis*.

Facilitated diffusion 4. *Dynamic equilibrium* is the process of passive transport in which proteins aid the passage of particles across the plasma membrane.

Pinocytosis 5. *Phagocytosis* is a form of endocytosis in which liquid droplets are taken in.

Answer the following questions.

6. How do active transport and passive transport differ? **In active transport, the cell uses energy to move particles from a region of lesser concentration to a region of greater concentration. In passive transport, the cell uses no energy.**

7. What is a selectively permeable membrane? **A selectively permeable membrane is one through which only certain types of molecules can permeate or pass.**

8. What is the role of carrier proteins in passive transport? **Carrier proteins change shape to allow certain ions and molecules to cross the plasma membrane.**

Explain how the terms in each set are related.

9. Diffusion, osmosis, facilitated diffusion
All are forms of passive transport.

10. Channel proteins, carrier proteins, gate proteins
All are types of transport proteins.

11. Phagocytosis, pinocytosis, receptor-aided endocytosis
All are forms of endocytosis in which the plasma membrane engulfs and then takes in substances from a cell's environment.

THE CELL AND ITS ENVIRONMENT

Interpreting and Applying Concepts

Refer to Figure 4–1 to answer the questions that follow.

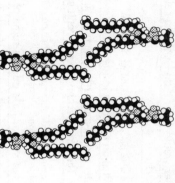

Phospholipid molecule

Figure 4–1

1. What is shown by Part A of the figure?
the polar head of the phospholipid

2. What is shown by Part B of the figure?
the nonpolar tails or fatty acids of the phospholipid

3. Describe how each part of the molecule behaves in water. **The polar head, labeled A, easily dissolves in water while the nonpolar tails, labeled B, do not dissolve in water.**

4. In the space below, sketch the arrangement of molecules, such as the one shown above, in the plasma membrane.

5. Does your sketch accurately depict all substances present in the plasma membrane? If not, identify the missing substances and explain their function.
No, the sketch lacks cholesterol molecules and membrane proteins. Cholesterol molecules make the membrane more rigid and keep the fatty acid tails apart. Membrane proteins regulate the flow of particles across the membrane, serve as enzymes, and act as recognition markers.

THE CELL AND ITS ENVIRONMENT

Interpreting and Applying Concepts continued

6. What is there about the structure of the plasma membrane that makes it freely permeable to water? What would happen to the cell if water could not freely pass in and out?

Because the polar heads of the phospholipids form the inner and outer boundaries of the plasma membrane, polar water molecules are attracted to the polar membrane. Water is able to pass through the membrane because it is such a small molecule. Without free passage of water throughout the membrane, the cell would be unable to obtain necessary nutrients and would die.

Study Figure 4–2, read the paragraph, and then answer the questions that follow.

A paramecium is a unicellular organism commonly found in ponds and slow-moving streams. These organisms contain structures called contractile vacuoles that are water-storage chambers. Excess water passing through the organism is stored in this structure. When the contractile vacuole is filled, it quickly contracts and pumps the water out of the cell.

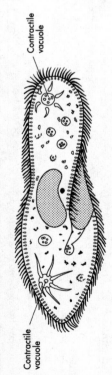

Contractile
vacuole

Contractile
vacuole

Figure 4–2

7. By what process does water enter a paramecium?

osmosis

8. What causes water to enter this unicellular organism?

differences in concentration of water between the cell and its watery

environment

9. Suppose a paramecium were removed from a pond and placed in a container of pure water. What would you observe?

Pure water would enter the cell by osmosis, causing the contractile vacuole to swell.

THE CELL AND ITS ENVIRONMENT

Interpreting and Applying Concepts continued

10. Do you think it is likely that the paramecium would survive in its new surroundings? Support your answer.

As long as the contractile vacuole was able to pump out excess water at the same rate that water entered the cell, it would survive. Most likely, however, more water would enter than leave the cell. As a result, the cell would eventually burst.

11. Suppose the paramecium were placed in a container of sea water. What would you observe?

Because the concentration of water inside the cell is greater than that outside the cell, water would move from the organism to its saltier sea water environment. The paramecium would eventually shrivel up and die.

12. How might you adapt the environment of the paramecium described in Question 9 to increase the chances of survival of the organism?

Answers will vary, but may include adding some of the original pond water to the container or placing a small amount of salt in the pure water.

Name Date Class

CHAPTER 5 ASSESSMENT

INSIDE THE CELL

Understanding Concepts

In the space at the left, write TRUE if the statement is true. If the statement is false, change the italicized word or phrase to make it true.

true 1. The sum of all chemical changes in cells is called *metabolism.*

Chloroplasts 2. *Lysosomes* are plastids that contain chlorophyll and other pigments needed for photosynthesis.

prokaryotes 3. Cells that do not have a membrane-bound nucleus are called *eukaryotes.*

true 4. *Symbiosis* refers to a relationship in which two organisms live closely together.

Explain how the terms in each set are related.

5. chromosomes, nucleus, chromatin
The nucleus contains chromatin which is made up of chromosomes.

6. microtubules, cytoskeleton, microfilaments
Microtubles and microfilaments are both structures of the cytoskeleton.

7. bacteria, yeasts, *Amoeba*
Bacteria, yeasts, and *Amoeba* are all unicellular organisms.

8. system, community, organism
Systems work together to form an organism and organisms work together to form communities.

In the space at the left, write the letter of the phrase from Column B that best matches the term in Column A.

	Column A	Column B
d	9. ribosomes	a. control center of a cell
e	10. mitochondria	b. delivery system in a eukaryotic cell
a	11. nucleus	c. vesicles formed from Golgi bodies
c	12. lysosomes	d. protein-making organelles
b	13. Golgi bodies	e. powerhouses of a cell

Copyright © by the Glencoe Division of Macmillan/McGraw-Hill School Publishing Company

Name Date Class

INSIDE THE CELL

Interpreting and Applying Concepts

Answer the following questions.

1. Suppose you were asked to classify an unidentified cell as either prokaryotic or eukaryotic. How would you go about doing this? **Viewing the specimen under a microscope would reveal whether or not the cell contains a membrane-bound nucleus. If so, it is a eukaryotic cell. If it lacks such a structure, it is a prokaryotic cell.**

2. Would careful study of an *Amoeba* reveal levels of organization within the organism? Explain your answer. **No, since an *Amoeba* is a unicellular organism, it lacks various levels of organization. The single cell performs all the functions necessary for life.**

3. To many scientists, the difference between prokaryotic cells and eukaryotic cells is more important than the difference between plant and animal cells. Do you agree with this assessment? Explain your answer. **Students will likely agree with this assessment because the first life forms were prokaryotes. Eukaryotes evolved from prokaryotes and represent more specialized life forms. Also, the majority of plant and animal cells are eukaryotes with similar cell structures.**

Figure 5–1 shows a human blood cell and a human nerve cell. Study the shapes of the cells and answer the questions that follow.

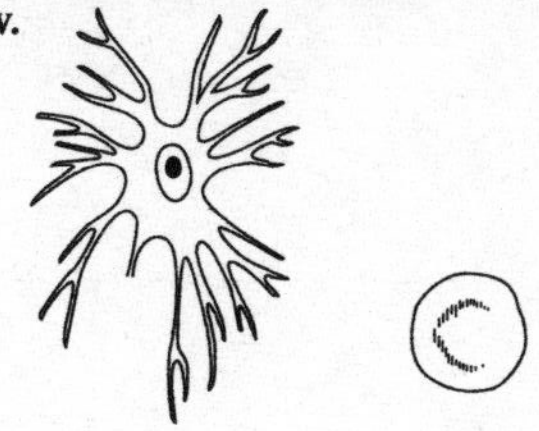

Figure 5–1 Cell A Cell B

4. Identify each cell. Describe the evidence that helped you identify each cell. **Cell A is a nerve cell. Its long, branched endings aid in transmitting impulses throughout the body. Cell B is a blood cell. Its rounded edges facilitate movement throughout the body's tiny blood vessels.**

5. Suppose the shapes of the cells illustrated in Figure 5–1 could be reversed. How might the change in shape affect the ability of the cell to do its job?
Neither type of cell would function with the other's shape. Students may explain that if blood cells had branched endings, they would become attached to the walls of blood vessels. If nerve cells had rounded edges, impulses would not be able to pass through easily.

Copyright © by the Glencoe Division of Macmillan/McGraw-Hill School Publishing Company

INSIDE THE CELL

Interpreting and Applying Concepts   continued

6. Suppose you were asked to classify an unidentified cell as either a plant or animal cell. How would you go about doing this? **Presence of a cell wall and chloroplasts would indicate that the cell is a plant cell. Lack of these structures provide evidence that the cell is an animal cell.**

7. Explain what is meant by the following statement: RNA is to ribosomes as DNA is to chromosomes. **Ribosomes are made up of the nucleic acid RNA and proteins, and chromosomes are made up of the nucleic acid DNA and proteins.**

Study Figure 5–2 and answer the questions that follow.

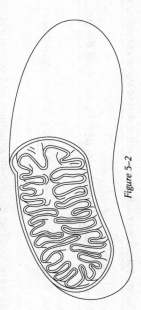

Figure 5–2

8. Identify the kind of cell organelle shown in Figure 5–2. Then explain its function. **The figure shows a mitochondrion. Mitochondria are powerhouses of cells where organic molecules are broken down to release energy.**

9. A human liver cell contains many more of the kind of organelle shown in Figure 5–2 than do various other types of cells. What does this indicate about the liver? Why? **The large number of mitochondria in a liver cell indicates that the liver does a great deal of work and therefore requires a great deal of energy. The more energy a cell needs, the more mitochondria it contains.**

10. How is the kind of organelle shown in Figure 5–2 similar to a chloroplast? **Both organelles change energy from one form to another. Mitochondria change the chemical energy stored in food to a form that the cell can use easily. Choloplasts change light energy to chemical energy.**

11. Do bacteria contain the structure shown in Figure 5–2? Explain your answer. **No; Bacteria are prokaryotes. As such, they do not contain mitochondria. Mitochondria are organelles present only in eukaryotes.**

THE FLOW OF ENERGY

Interpreting and Applying Concepts

Figure 6–1 illustrates the structure of a chloroplast. Use the figure to answer the questions that follow. Write your answers using complete sentences. Use the names of Parts A, B, and C in your answers.

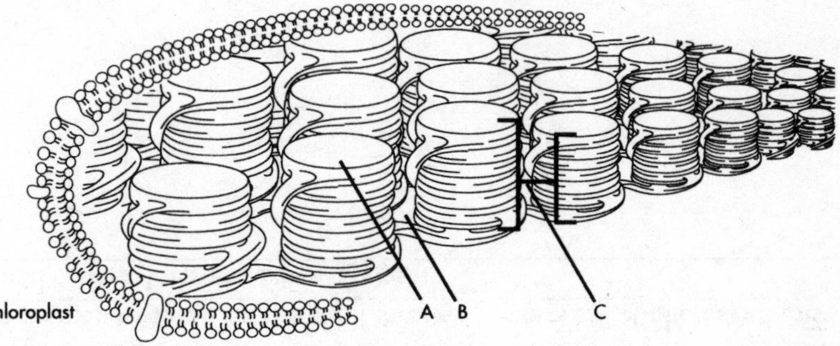

Chloroplast

Figure 6–1

1. Explain the relationship that exists between Part A and Part C of the figure.
Part C, or grana, is actually a stack of thylakoid membranes labeled A in the figure.

2. In which part of the figure would you most likely find 3-carbon sugars? Explain your answer.
Since the Calvin cycle takes place in the stroma, it is likely that Part B of the figure would contain 3-carbon sugars.

3. Of the labeled structures shown in the figure, which has a structure similar to that of the cell membrane? Explain how they are alike.
Part A of the figure. The thylakoid membrane is similar in structure to a cell membrane. Both have a fluid mosaic structure with proteins embedded in their bilayers.

4. Predict the color of Part A of the figure. Provide reasons for your prediction.
The pigment chlorophyll is green in color. Chlorophyll molecules are located on the thylakoid membranes inside a chloroplast. Therefore, it is likely that Part A of the figure would be green in color.

5. The structure of a hemoglobin molecule is similar to that of a chlorophyll molecule. The greatest difference between these substances is the presence of an atom of iron in the center of a hemoglobin molecule rather than the magnesium atom found in a chlorophyll molecule. What might this similarity indicate?
It is possible that plants and animals evolved from a common ancestor.

Copyright © by the Glencoe Division of Macmillan/McGraw-Hill School Publishing Company

---

CHAPTER 6 ✿ ASSESSMENT

THE FLOW OF ENERGY

Understanding Concepts

In the space at the left, write TRUE if the statement is true. If the statement is false, change the italicized word or phrase to make it true.

true 1. The *wavelength* of any form of radiant energy is the distance between one wave crest and the next.

Endergonic 2. *Exergonic* reactions are chemical reactions that require free energy.

light energy 3. Energy for the process of photosynthesis comes from *chemical energy* absorbed by chlorophyll.

true 4. In a chloroplast, sugar is synthesized in the *stroma*.

visible spectrum 5. The range of colors that make up white light is known as the *absorption spectrum*.

Explain how the terms in each set are related.

6. lactic acid fermentation, alcoholic fermentation, anaerobic processes
Lactic acid fermentation and alcoholic fermentation are two anaerobic processes.

7. chlorophylls, photosynthesis, carotenoids
Chlorophylls and carotenoids interact to absorb the light energy necessary for photosynthesis.

8. aerobic respiration, glucose, oxygen
Glucose and oxygen are two reactants of aerobic respiration.

9. carbon dioxide, Calvin cycle, sugar, hydrogen ions
The reactions of the Calvin cycle convert carbon dioxide and hydrogen ions to sugars.

10. exergonic reactions, ATP, endergonic reactions
In cells, ATP is the main energy link between exergonic and endergonic reactions.

Answer the following question.

11. What are two ways in which photosynthesis and respiration are related?
The waste products of one process are the raw materials for the other. Photosynthesis stores energy in sugar molecules and respiration breaks down those molecules to release energy for cell functions.

Copyright © by the Glencoe Division of Macmillan/McGraw-Hill School Publishing Company

131

Interpreting and Applying Concepts continued

Refer to Figure 6–2. Answer the questions that follow in complete sentences.

Figure 6–2 **Where Photosynthesis Occurs on Earth**

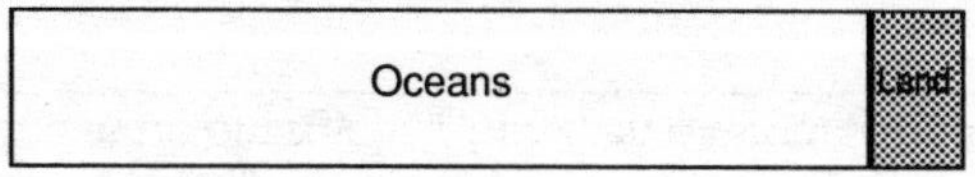

Oceans	Land

6. Based on the graph, estimate what percent of all photosynthesis takes place in Earth's oceans. **The graph shows that about 90 percent of all photosynthesis takes place in the oceans.**

7. Explain the danger the world faces if oceans are polluted to the point where populations of photosynthetic organisms are seriously affected. **The loss of these populations would result in a reduction of the oxygen content of the atmosphere. Animals, including humans, rely on that oxygen for life processes. In addition, most other ocean life would die off.**

8. Most cells rely on ATP as a ready energy source to carry out life processes. How do cells get more ATP to replace what has been used? **Cells break down molecules to release energy that is used to make more ATP from ADP and inorganic phosphate.**

9. What is fermentation? **Fermentation in cells is an anaerobic process in which energy is produced in the absence of oxygen.**

10. Describe the type of fermentation that could occur in your cells. **Lactic acid fermentation can occur in muscles of the human body. In this process, enzymes break down a glucose molecule into two lactic acid molecules and transfer energy to ATP.**

11. Compare the products of the type of fermentation described in Question 10 with those of fermentation that occurs in yeast cells. **Lactic acid fermentation produces lactic acid while alcoholic fermentation produces alcohol and carbon dioxide. Both processes also produce two molecules (net) of ATP.**

Interpreting and Applying Concepts continued

12. Yeast cells are living organisms that are dependent on definite temperature ranges. Yeast is dormant at cold temperatures, but begins to activate at about 50°F and is most active between 80°F and 100°F. Yeast cells begin to die at temperatures around 120°F. A bread recipe calls for dissolving yeast in warm (85°F) water. Predict the effect each of the following actions would have on the dough-rising process.

a. The baker dissolves the yeast in lukewarm (60°F) water. **The rising process would probably occur, but it would take longer because the warm water would not activate the yeast as quickly as hot water.**

b. The baker dissolves the yeast in boiling water. **The rising process would probably not take place at all because the yeast cells would be killed by the boiling water.**

13. In an effort to reduce energy needs, the alternative fuel gasohol was developed. This mixture consists of 90 percent gasoline and 10 percent ethanol. Explain how yeast could be used to make gasohol. **The alcoholic fermentation of yeast produces carbon dioxide and ethanol. The ethanol produced could be combined with gasoline to make gasohol.**

14. In a particular experiment, scientists used water molecules that contained the radioactive isotope oxygen-18. What do you suppose would occur if this water were added to the soil of a potted green plant? **The oxygen released by the plant through the process of photosynthesis would contain the radioactive oxygen.**

CELLULAR REPRODUCTION

Interpreting and Applying Concepts

Use Figure 7-1 showing the phases of the cell cycle to answer the questions.

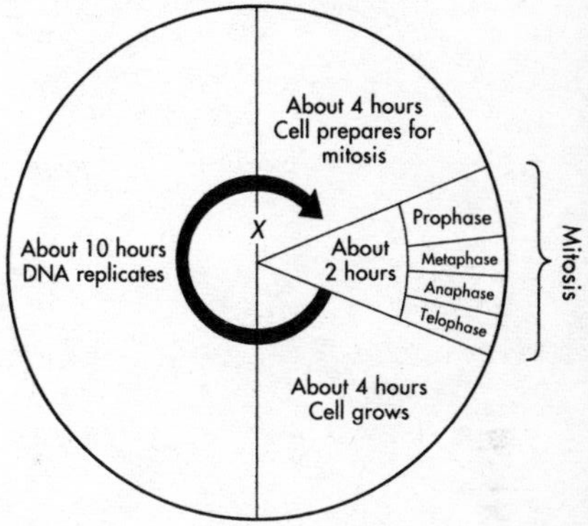

Figure 7-1   A 20-Hour Cell Cycle

1. What part of the cycle does X represent?
   interphase

2. What is the most important process to occur during that part of the cycle?
   chromosome (DNA) replication

3. About what percentage of the cell cycle is taken up by mitosis?
   about 10%

4. Suppose a drug that prevents proper development of cell structures in the last few hours of interphase is used. What would be the result? **Mitosis cannot occur; the cell usually dies.**

When meiosis I occurs following fertilization in an organism with four chromosomes, two alternate arrangements of the chromosomes are possible. Refer to Figure 7-2 and answer the question that follows.

Figure 7-2

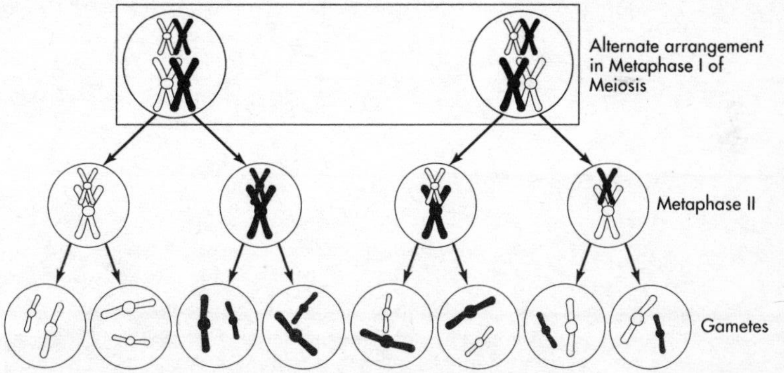

5. Complete the figure. Draw the chromosomes in each cell in metaphase II to show the possible arrangements of the chromosomes. Then draw the chromosomes in the gametes to show the possible combinations of chromosomes. (You may use colored pencils.)

Copyright © by the Glencoe Division of Macmillan/McGraw-Hill School Publishing Company

---

CHAPTER 7 ❧ ASSESSMENT
CELLULAR REPRODUCTION

Understanding Concepts

In the space at the left, write the letter of the word or phrase that best completes the statement or answers the question.

a 1. In your body, cells that could *not* be undergoing cell division are
   a. nerve cells.   b. skin cells.   c. red blood cells.   d. liver cells.

c 2. Prokaryotes produce daughter cells by the process of
   a. meiosis I.   b. meiosis II.   c. binary fission.   d. mitosis.

d 3. In organisms that reproduce by sexual reproduction, the 2n number of chromosomes is restored by
   a. cell division.   b. binary fission.   c. spore formation.   d. fertilization.

b 4. The two processes that provide for variation and genetic recombination among offspring are
   a. fertilization and mitosis.   c. cell division and mitosis.
   b. fertilization and meiosis.   d. cell division and meiosis.

c 5. Pasteur's work led directly to the
   a. cell theory.   c. theory of biogenesis.
   b. theory of continuance.   d. theory of spontaneous generation.

a 6. You know you are observing plant cells undergoing mitosis if you observe
   a. cell plates.   c. tetrads.
   b. centrioles.   d. endoplasmic reticulum.

b 7. Meiosis in humans results in
   a. one sperm cell.   c. four identical sperm cells.
   b. one egg cell.   d. four identical egg cells.

a 8. In plants and some algae and fungi, meiosis results in
   a. haploid spores.   b. gametes.   c. microtubules.   d. flagella.

In the space at the left, write TRUE if the statement is true. If the statement is false, change the italicized word or phrase to make it true.

zygote   9. The union of a sperm cell and an egg cell results in a *gamete.*

true   10. The region at which the two strands of a chromosome are joined is the *centromere.*

mitosis   11. The process that ensures that each daughter cell has the same number and kind of chromosomes as the parent cell is *meiosis.*

Copyright © by the Glencoe Division of Macmillan/McGraw-Hill School Publishing Company

CELLULAR REPRODUCTION

Interpreting and Applying Concepts *continued*

The diagrams in Figure 7–3 show cells from the same eukaryotic animal cell in various stages of mitosis. Use the diagrams to answer the questions.

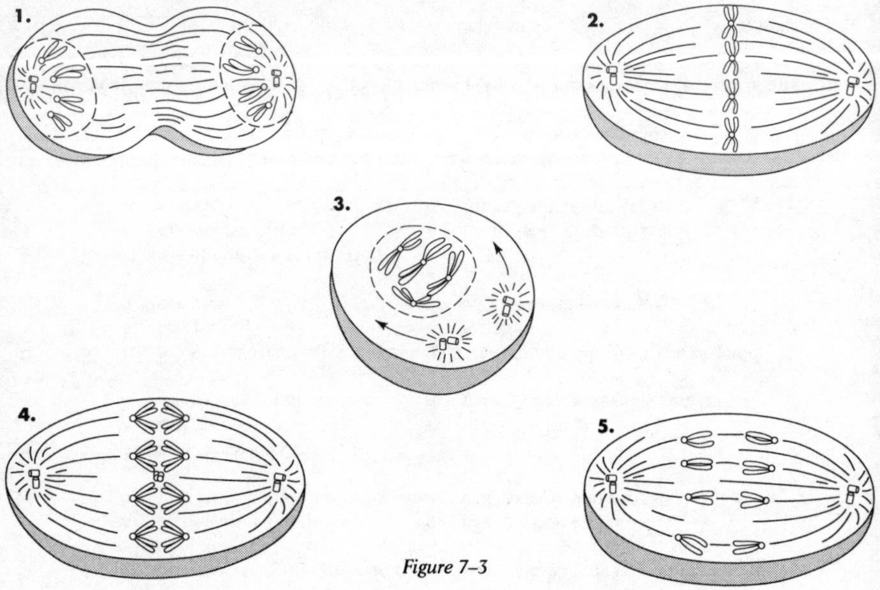

1.

2.

3.

4.

5.

Figure 7–3

6. What is the chromosome number for the cells in the diagram? _____ **4**

7. What is the structure labeled *X* in diagram 2? **the spindle**

8. Which diagram correctly represents a cell in anaphase? **5**

9. How many chromatids are shown in diagram 2? **8**

10. Which diagram shows a stage of mitosis incorrectly? **4**

11. Which structure indicates that you are observing mitosis in an animal cell? **centriole**

12. List the proper order for the four *correct* diagrams. **3,2,5,1**

HEREDITY

Interpreting and Applying Concepts

When a white, flat squash (*WWFF*) is crossed with a yellow, round squash (*wwff*), all the offspring are white and flat (*WwFf*). Suppose two of these offspring are crossed. The partial results in the F$_2$ generation are shown in the Punnett square in Figure 8-1. Study Figure 8–1 and then answer the questions that follow.

Figure 8–1

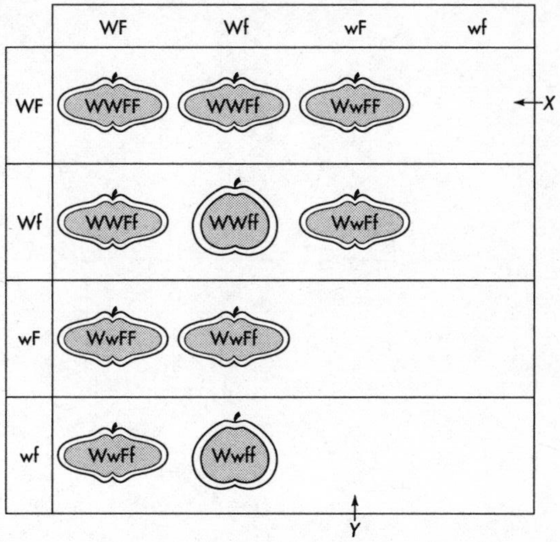

1. How many white, round squash would be indicated by the complete Punnett square?

   3

2. Draw the squash that would appear in the square labeled *x*. Be sure to follow the pattern for shape and color indicated.

3. What is the genotype of the squash in the square labeled *x*?

   WwFf

4. Draw the squash that would appear in the square labeled *y*.

5. What is the phenotype of the squash in the square labeled *y*.

   yellow, flat

6. Give the ratio for all the phenotypes that would be indicated by the complete Punnett square. (Include traits in answer.)

   9 white flat : 3 white round : 3 yellow flat : 1 yellow round

Copyright © by the Glencoe Division of Macmillan/McGraw-Hill School Publishing Company

---

CHAPTER 8 ❧ ASSESSMENT

HEREDITY

Understanding Concepts

Write the letter of the word or phrase that best completes the statement or answers the question.

1. When a pure tall pea plant is crossed with a pure short pea plant, the ___**dominant**___ trait appears in the F$_1$ generation.

2. The phenotypic ratio obtained when two organisms that are heterozygous for two different traits are crossed is ___**9 : 3 : 3 : 1**___.

3. The genotypic ratio obtained when an *Rr* organism is crossed with an *rr* organism is ___**1 Rr : 1 rr**___. (Include letters.)

4. When two pea plants that are heterozygous for green pods are crossed, the genotypic ratio that results is ___**1GG : 2Gg : 1gg**___. (Include letters.)

In the space at the left, write the term in parentheses that correctly completes the statement.

closer to   5. According to the laws of probability, the larger the number of trials, the (*closer to, farther from*) the expected ratio the results will be.

multiple genes   6. Human skin color, which shows continuous variation, appears to be inherited through (*multiple genes, modifier genes*).

heterozygous   7. An organism with two unlike alleles for the same trait is (*homozygous, heterozygous*).

can   8. The environment in which an organism develops (*can, cannot*) affect the expression of its genetic traits.

In the space at the left, write the letter of the phrase from Column B that best matches the term in Column A.

Column A		Column B
g	9. autosomes	a. An individual with two similar alleles for the same trait
f	10. codominance	b. The science of heredity
b	11. genetics	c. Genes for different traits segregate without regard for each other
a	12. homozygous	d. The physical appearance of an organism
e	13. incomplete dominance	e. Two alleles result in an intermediate third phenotype
d	14. phenotype	f. Two alleles expressed equally in the offspring phenotype
c	15. principle of independent assortment	g. Chromosomes not involved in determining the sex of an individual

Copyright © by the Glencoe Division of Macmillan/McGraw-Hill School Publishing Company

135

HEREDITY

Interpreting and Applying Concepts continued

Assume that a mother with blood type *A* has a child of blood type *O*. Answer the following questions.

7. What is the genotype of the mother? How do you know? I^Ai. **For the baby to have type O blood, the mother would have to have at least one *i* allele.**

8. What blood types could the father be? **AB**

9. If the father was type *A* would he be I^AI^A or I^Ai? **I^Ai**

10. If the father was type *B*, would he be I^BI^B or I^Bi? **I^Bi**

11. Must the father be type *O*? Why or why not? **No. He could be I^Ai or I^Bi. Then he could by chance contribute the *i* allele for blood type to the child.**

Answer the following questions.

12. Draw and label a Punnett square containing the human sex chromosomes to show the inheritance of maleness and femaleness in humans. (Use the appropriate letters for the chromosomes.)

		Male	
		X	**Y**
Female	**X**	XX	XY
	X	XX	XY

13. Use the data from the Punnett square to explain why approximately half of all infants born are male and half are female. **Half of the offspring have the XX chromosome pattern. This XX pattern is found in females. The other half of the offspring have the XY chromosome pattern. The XY pattern is found in males.**

14. Use the data from the Punnett square to help explain why sex-linked traits, such as color blindness, may appear occasionally in men but very rarely in women. **The gene for color blindness appears on the X chromosome. However, the Y chromosome has no gene related to color vision. Even though the gene is recessive, a male with only one gene for color blindness on his X chromosome would be color-blind. A woman would have to have a recessive gene for color blindness or both of her X chromosomes to be color-blind.**

Copyright © by the Glencoe Division of Macmillan/McGraw-Hill School Publishing Company

CHAPTER 8 ASSESSMENT

31

CHEMISTRY OF THE GENE

Interpreting and Applying Concepts

Read the paragraph and answer the questions that follow.

In 1958, the replication model was verified by the experiments of Meselson and Stahl. They grew *E. coli* bacteria in a medium containing N^{15} (heavy form of nitrogen). After many generations, all their DNA molecules contained only N^{15}. Then these bacteria were grown on a medium containing N^{14} (ordinary light nitrogen). After one cell division, the DNA in the bacteria was halfway in weight between N^{15} and N^{14}. Each new DNA molecule had one strand containing N^{15} and one strand containing N^{14}. Finally, these bacteria divided again on the N^{14} medium.

1. Complete Figure 9-1 to show the types of DNA in the four cells that developed in the second generation.

Figure 9-1

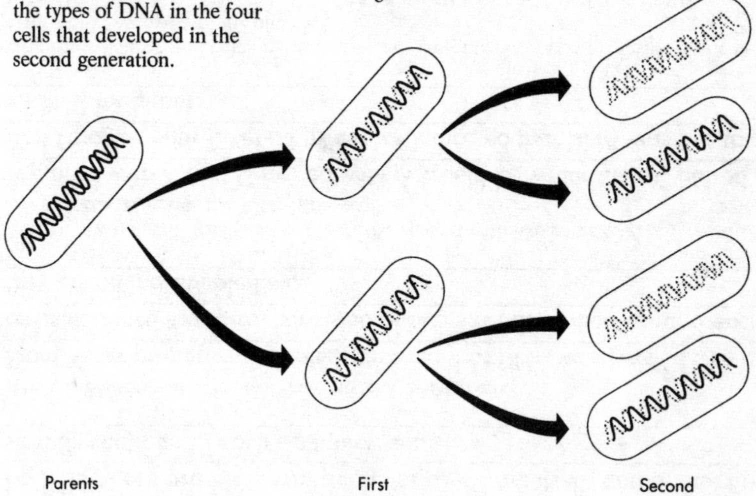

Parents   First generation   Second generation

2. Explain your answer to Exercise 1. **Each DNA molecule from the first generation replicates. The N^{14} strand replicates an N^{14} strand from the N^{14} medium. The N^{15} strand can replicate only a new N^{14} strand from the N^{14} medium.**

_____

_____

Copyright © by the Glencoe Division of Macmillan/McGraw-Hill School Publishing Company

---

CHAPTER 9 ❧ ASSESSMENT

CHEMISTRY OF THE GENE

Understanding Concepts

In the space at the left, write the letter of the word or phrase that best completes the statement or answers the question.

d 1. Each of these scientists was involved in identifying the transforming principle *except*
   a. Griffith.   **b.** Avery.   **c.** Hershey and Chase.   **d.** Watson and Crick.

a 2. The genetic code in DNA depends on the order of the
   a. nucleotides.   **b.** pyrimidines.   **c.** major grooves.   **d.** phosphate groups.

d 3. Protein synthesis involves the use of
   a. 6 nucleotides.   **c.** 20 polypeptides.
   b. 16 amino acids.   **d.** 64 codons.

c 4. A nucleotide consists of each of the following *except* a(n)
   a. deoxyribose sugar.   **c.** amino acid.
   b. phosphate group.   **d.** nitrogen-containing base.

c 5. In the ladder-like model of DNA, the uprights consist of
   a. nucleotides.   **c.** sugar and phosphate groups.
   b. nitrogen bases.   **d.** ATP.

c 6. In a eukaryote, replication of DNA *always* begins
   a. at one end.   **c.** at several places along the DNA.
   b. between two codons.   **d.** within one of the nucleotides.

b 7. During replication of DNA, a single mistake in a nucleotide
   a. is always fatal.   **c.** occurs in 1 of 1000 nucleotides.
   b. is usually self-repairing.   **d.** prevents variation in genes.

d 8. One segment of nucleotides along a DNA molecule forms a
   a. purine.   **b.** pyrimidine.   **c.** phage.   **d.** gene.

In the space at the left, write the term in parentheses that best completes the statement.

codon 9. In DNA, each set of three nitrogen bases representing an amino acid is a (*nucleotide, codon*).

a complement 10. After replication, each new strand of DNA, compared with one of the two parent strands, is (*a complement, identical*).

polypeptides 11. In cells, the long chains formed of hundreds of amino acids are (*polypeptides, phosphates*).

ATACG 12. If one strand of DNA has the nitrogen bases TATGC, the other strand has the bases (*TATGC, ATACG*).

Copyright © by the Glencoe Division of Macmillan/McGraw-Hill School Publishing Company

CHEMISTRY OF THE GENE

Interpreting and Applying Concepts   continued

Study the codons of DNA in Table 9–1 to answer the questions.

Table 9–1

First Base in Codon	Second Base in Codon				Third Base in Codon
	A	G	T	C	
A	phenylalanine	serine	tyrosine	cysteine	A
	phenylalanine	serine	tyrosine	cysteine	G
	leucine	serine	stop	stop	T
	leucine	serine	stop	tryptophan	C
G	leucine	proline	histidine	arginine	A
	leucine	proline	histidine	arginine	G
	leucine	proline	glutamine	arginine	T
	leucine	proline	glutamine	arginine	C
T	isoleucine	threonine	asparagine	serine	A
	isoleucine	threonine	asparagine	serine	G
	isoleucine	threonine	lysine	arginine	T
	methionine	threonine	lysine	arginine	C
C	valine	alanine	aspartate	glycine	A
	valine	alanine	aspartate	glycine	G
	valine	alanine	glutamate	glycine	T
	valine	alanine	glutamate	glycine	C

3. What does each codon represent? __an amino acid__

4. What are the three letters that make up each stop codon? __ATT, ATC, ACT__

5. What two amino acids are represented by only one codon each? __methionine and tryptophan__

6. In codons for the same amino acids, the first two nitrogen bases are usually the same. Which amino acids form an exception to this rule? __leucine and arginine__

7. Codon TAC has a dual function. One function is to signal the start of protein synthesis. What is the other function? __It codes for the amino acid methionine.__

CHEMISTRY OF THE GENE

Interpreting and Applying Concepts   continued

Read the paragraph and answer the questions that follow.

In building their model of DNA, Watson and Crick built on the work of other scientists who had studied various aspects of the structure of DNA. However, Watson and Crick had to answer several major questions about the structure of DNA in order to develop their model. This now-famous model accounted for the apparent uniform width of DNA and explained a great mystery: How could DNA copy itself?

8. Linus Pauling proposed a three-stranded model of DNA. To Watson and Crick, the width of the helix suggested that DNA was two-stranded. Why else did they reason it was two stranded? __Each body cell produces two daughter cells from itself, and one of the two strands could go to each daughter cell.__

9. Why did Watson and Crick call their model a double helix? __From X-ray photographs, they reasoned that DNA was twisted into a spiral, or helix. Since the spiral consisted of two strands wound around each other, they called it a double helix.__

10. At first, Watson and Crick believed that bases paired with like bases — A with A, C with C, and so forth. Why was this idea disproved? __A paired with A and G paired with G would be wider than C paired with C and T paired with T. Yet all evidence indicated that DNA was the same width throughout.__

11. Originally, Watson and Crick placed the sugar-phosphate chain on the inside of the molecule. Why did they move it to the outside? __With bases on the outside, the DNA would not be uniform width throughout--which all evidence indicated. Also, the weak hydrogen bonds between the nitrogen bases could be broken easily. The bonds between the sugar and phosphate portions of the nucleotides are much stronger.__

12. In 1953, Watson and Crick surprised the scientific community when they published a report of their model of DNA in the journal *Nature*. Why has this model become a cornerstone of modern molecular biology? __Their model of DNA suggested a mechanism for DNA replication, essential to the modern concept of genetics.__

CHAPTER 10 ✹ ASSESSMENT

FROM GENES TO PROTEINS

Understanding Concepts

In the space at the left, write the term in parentheses that best completes the statement.

uracil — 1. RNA, unlike DNA, contains the nitrogen base (*uracil, thymine*).

transcription — 2. The process of transferring the genetic information from DNA to RNA is called (*translation, transcription*).

point — 3. A gene mutation that affects only one amino acid is a (*point, chromosome*) mutation.

introns — 4. Nucleotide sequences that do not code for amino acids are (*exons, introns*).

ribosomes — 5. During protein synthesis, the (*codons, ribosomes*) move along the mRNA strand.

recombinant — 6. In bacteria, DNA that has been altered by insertion of a foreign gene is called (*recombinant, template*) DNA.

one percent — 7. Scientists estimate that approximately (*one percent, 90 percent*) of DNA in eukaryotic cells codes for polypeptides.

sex — 8. Mutations may affect an entire population of organisms if they occur in (*body, sex*) cells.

jumping genes — 9. Barbara McClintock received a Nobel Prize about 40 years after her discovery of (*pseudogenes, jumping genes*).

malignant — 10. Abnormal cells that can migrate from one organ to another are (*gene-cloned, malignant*) cells.

transgenic — 11. Plants that contain genetic materials from two species are (*transformed, transgenic*) plants.

single — 12. RNA is usually composed of a (*single, double*) chain of nucleotides.

transfer — 13. In protein synthesis, amino acids are brought to the ribosome by (*transfer, messenger*) RNA.

ribose — 14. The sugar present in RNA is (*ribose, deoxyribose*).

In the space at the left, write the letter of the phrase from Column B that best matches the term in Column A.

	Column A		Column B
a	15. anticodon	a.	three bases at one end of a tRNA molecule
e	16. clones	b.	changes in the genes
b	17. mutations	c.	genes that mutate and cause cancer
c	18. oncogenes	d.	small circles of DNA in bacteria
d	19. plasmids	e.	genetically identical copies of organisms

FROM GENES TO PROTEINS

Interpreting and Applying Concepts

Study Figure 10–1, in which four stages of protein synthesis, labeled 1–4, are shown out of order, and then answer the following questions.

Figure 10–1

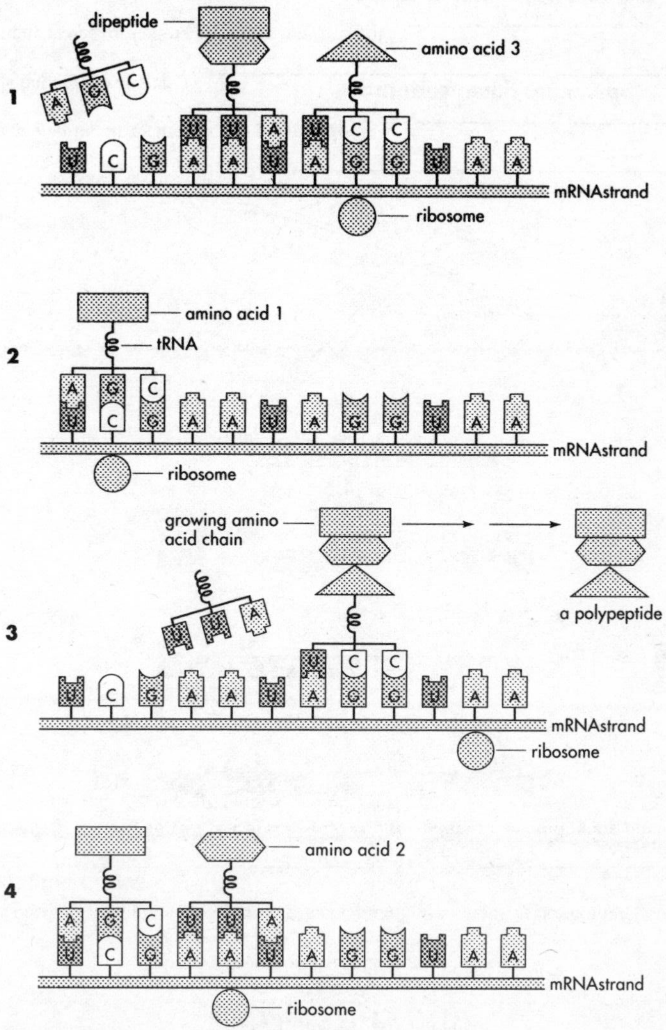

139

FROM GENES TO PROTEINS

Interpreting and Applying Concepts continued

1. Use the numbers of the stages to list them in proper order. **2,4,1,3**

2. How many different tRNA molecules are shown in the figure? Explain your answer. **Three; each amino acid involves only one tRNA molecule, even though each is shown several times.**

3. How many different mRNA strands are shown in the figure? Why do you say so? **One; the same mRNA strand is shown in each of the diagrams.**

4. Where is a stop codon located? How do you know? **The three nitrogen bases (UAA) at the right end of the mRNA strand form a stop codon. The chain of three amino acids is released to form a polypeptide.**

5. What process had to precede the portion of protein synthesis shown in the figure? Explain why. **Transcription; the genetic information present in the DNA in the nucleus had to pass to single-stranded RNA in the cytoplasm.**

Study Figure 10–2, which shows four kinds of chromosome mutations, and then answer the following questions.

Diagram 1

Diagram 2

Diagram 3

Diagram 4

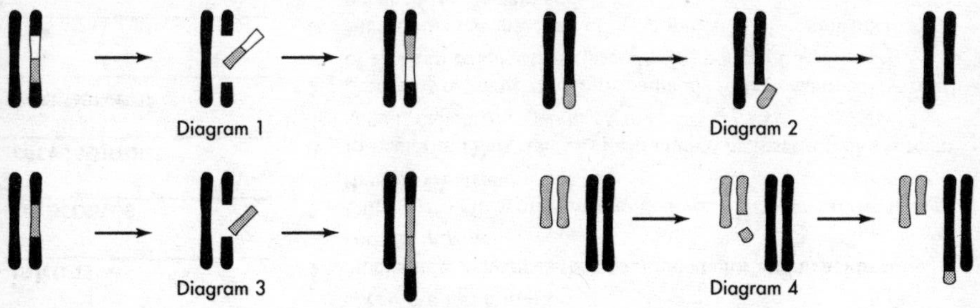

Figure 10–2

6. What type of mutation is shown in diagram 1? **inversion**

7. What type of mutation is shown in diagram 2? **deletion**

8. What type of mutation is shown in diagram 3? **duplication**

9. Diagram 4 shows translocation, another type of chromosome mutation. Describe what happens during translocation. **A segment of one chromosome becomes attached to another chromosome.**

10. What types of chromosomal mutations involve change in one chromosome of a pair? **deletion, inversion, duplication**

FROM GENES TO PROTEINS

Interpreting and Applying Concepts continued

11. Are the chromosome pairs involved in translocation homologous or nonhomologous? **nonhomologous**

Study Figure 10–3, which shows RNA processing, and then answer the following questions.

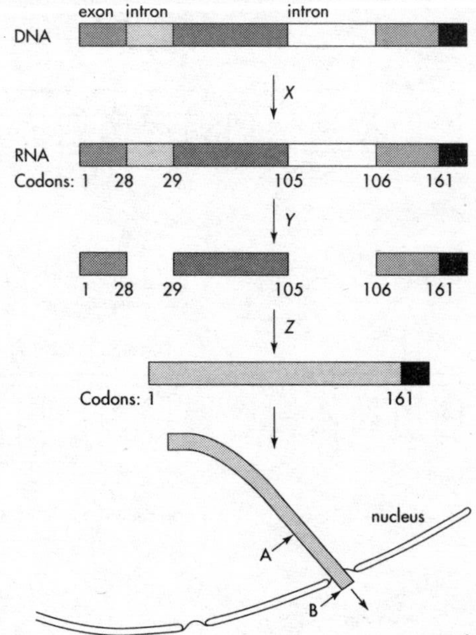

Figure 10–3

12. What process is represented at step X? **transcription**

13. How many amino acids long is this polypeptide? **161**

14. What is happening at step Y? **introns being removed**

15. How many exons, or coding segments, are present? **3**

16. What is happening at step Z? **exons being joined together**

17. What kind of RNA is shown at A? **messenger RNA**

18. Describe what is happening at B. **mRNA is leaving the nucleus and entering the cytoplasm.**

CHAPTER 11 ❧ ASSESSMENT

HUMAN GENETICS

Understanding Concepts

In the space at the left, write the term in parentheses that best completes the statement.

genetic counseling   1. The process by which parents are given information about chances of their unborn child having an inherited disorder is called (*genetic counseling, gene therapy*).

trisomy   2. The presence of three of one kind of chromosome in a cell is called (*trisomy, nondisjunction*).

homozygous recessive   3. Most human genetic disorders result from a (*homozygous recessive, heterozygous dominant*) genotype.

mutant   4. Cystic fibrosis, Huntington's disease, and sickle-cell anemia are the result of (*missing, mutant*) alleles.

In the space at the left, write the letter of the word or phrase that best completes the statement or answers the question.

b   5. A fetus is viewed directly through an endoscope in
   a. amniocentesis.
   b. fetoscopy.
   c. echograms.
   d. chorionic villus biopsy.

b   6. In parts of Africa where malaria is present, blacks who are heterozygous for sickle-cell hemoglobin tend to live longer than those with other genotypes. These individuals
   a. suffer from the disease.
   b. have heterozygous superiority.
   c. often die of malaria.
   d. have severe pain.

c   7. Huntington's disease is caused by
   a. an extra chromosome.
   b. two recessive genes.
   c. a single dominant gene.
   d. a missing chromosome.

a   8. An autosomal recessive disorder in which degeneration of the nervous system begins before the end of a baby's first year is
   a. Tay-Sachs disease.
   b. Down syndrome.
   c. diabetes mellitus.
   d. leukemia.

c   9. Each of the following is related to cystic fibrosis *except*
   a. abnormal cell membrane protein.
   b. most common inherited disorder among whites.
   c. deteriorating brain cells.
   d. accumulation of mucus in lungs.

b   10. If one parent has a disorder caused by a dominant gene, the chance of the offspring having the disorder
   a. is 0 percent.
   b. is 50 percent.
   c. is 100 percent.
   d. cannot be determined.

HUMAN GENETICS

Interpreting and Applying Concepts

One type of antigen in all human blood is the Rh factor. It is present in about 85 percent of the people in the United States, who are classified as Rh-positive (Rh+). People whose blood does not contain the Rh factor are classified as Rh negative (Rh−). Study Table 11–1, which indicates the inheritance of the Rh factor within 98 families. Then answer the questions that follow.

Table 11–1

Parent	Number of families	Children: Rh+	Children: Rh−	Percentage Rh-negative
Rh+ × Rh+	69	217	15	6.5
Rh+ × Rh−	21	55	24	30.4
Rh− × Rh−	8	0	37	100.0

1. What percentage of children born to two Rh-positive parents were Rh-negative?   **6.4%**

2. How many Rh-positive children were born to families in which one parent was Rh-positive and the other Rh-negative?   **55**

3. What percentage of children born to families in which both parents were Rh-negative were Rh-positive?   **0%**

4. Is the presence of the Rh factor a dominant or a recessive trait? Use the table to explain. **Since 100 percent of children born to two Rh-negative parents were Rh-negative, the absence of the factor must be recessive. Hence the presence of the factor must be dominant.**

5. In families having one Rh-positive parent and one Rh-negative parent, was the Rh-positive parent homozygous or heterozygous? How do you know?   **heterozygous; If the Rh-positive parent were homozygous, there could have been no Rh-negative children.**

141

HUMAN GENETICS

Interpreting and Applying Concepts   continued

Read the paragraph and then answer the following questions.

Some individuals contain a gene that causes blue sclera, a condition in which the whites of the eyes appear bluish. Anyone who possesses the gene, whether homozygous or heterozygous, is expected to show blue sclera. But it has been found that only about nine out of ten people with this gene show blue sclera. Among those showing the phenotype, the blue color ranges from pale whitish blue to dark blackish blue.

Scientists study this condition for penetrance and expressivity. Penetrance is the percentage of individuals carrying the gene that actually show it. Expressivity is the way the phenotype is expressed.

6. Is the gene for blue sclera dominant or recessive? How do you know? **dominant; Since heterozygotes, as well as homozygotes, show the trait, it must be dominant.**

7. If two individuals homozygous for blue sclera marry, what percentage of their offspring would you expect to have the gene? **100 percent**

8. What percentage would you expect to actually show blue sclera? Why do you say so? **90 percent; Penetrance of the gene for blue sclera is 90 percent.**

9. What is the expressivity of the phenotype for blue sclera? **Among people with blue sclera, the whites of the eyes range from pale whitish blue to deep blackish blue.**

Study Figure 11–1. Then read the paragraph and answer the questions that follow.

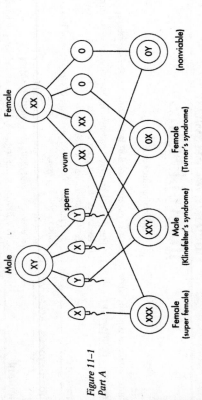

Figure 11–1
Part A

HUMAN GENETICS

Interpreting and Applying Concepts   continued

Figure 11–1
Part B

Figure 11–1 shows the result of nondisjunction, a failure of paired chromosomes to separate, in sex chromosomes of male and females. Part A shows the result of nondisjunction in females; Part B shows the result in males.

An individual with Turner's syndrome is a female, but does not develop normally and is sterile. She may be mildly retarded. An individual with Klinefelter's syndrome may appear to be a normal male, but is probably sterile and may show some mental retardation. Although a super female has an extra X chromosome, she does not have super ability and may be sterile.

10. What is the sex chromosome pattern of an individual with Klinefelter's syndrome? **XXY**

11. What is the sex chromosome pattern of an individual with Turner's syndrome? **OX**

12. What is the sex chromosome pattern of the four possible egg cells resulting from nondisjunction in females? **two XX and two O**

13. What is the sex chromosome pattern of the four possible sperm cells resulting from nondisjunction in males? **two XY and two O**

14. What is the meaning of the O in the diagrams? **The O indicates the lack of a sex chromosome, either X or Y.**

142

CHAPTER 12 ASSESSMENT

EVOLUTION

Understanding Concepts

In the space at the left, write the letter of the word or phrase that best completes the statement or answers the question.

__b__ 1. The forelegs of a cat and bat are examples of
   a. comparative biochemistry.   c. analogous structures.
   b. homologous structures.   d. comparative embryology.

__d__ 2. The similarity among the blood proteins of all the mammals may be taken as evidence for evolutionary relationships based upon
   a. comparative anatomy.   c. geographic distribution.
   b. comparative embryology.   d. comparative biochemistry.

__a__ 3. Which conclusion may be drawn when comparing fossils found in previously undisturbed strata of sedimentary rock?
   a. Fossils in the upper strata are younger than those in the lower strata.
   b. Fossils in the upper strata are older than those in the lower strata.
   c. Fossils in the upper strata are generally less complex than those in the lower strata.
   d. There are no fossils in the upper strata that resemble those in the lower strata.

__c__ 4. Which of the following was *not* part of Darwin's proposals?
   a. There is overpopulation in nature.
   b. There is variation in the individuals of the population.
   c. The variations are caused by mutations.
   d. There is natural selection.

Answer the following questions.

5. Why is the fossil evidence incomplete? **The fossil evidence is incomplete because fossil formation rarely occurred. Soft tissue decomposed and did not provide suitable source for fossilization. The conditions were not always conducive to the formation of fossils.**

6. What types of organisms have humans selectively bred? **Humans have selectively bred farm animals, agricultural crops, and household pets.**

7. Do bacteria exposed to penicillin mutate in response to the penicillin or do they already possess the mutation? **The bacteria that already possess the mutation that confers penicillin resistance are the bacteria that will survive when exposed to the penicillin. Mutations do not occur in response to environmental changes.**

8. What four elements necessary for life were contained in the gases that made up Earth's early atmosphere? **The four elements were carbon, hydrogen, oxygen, and nitrogen.**

Copyright © by the Glencoe Division of Macmillan/McGraw-Hill School Publishing Company

EVOLUTION

Interpreting and Applying Concepts

Answer the following questions in complete sentences.

1. Darwin calculated that, if a single breeding pair of elephants reproduced and all their offspring survived and reproduced, in 750 years there would be a standing population of 19 million elephants. In Africa and Asia, where elephants live in the wild, the total number of elephants is currently less than one million. Explain this in terms of natural selection.
According to Darwin there is a struggle for existence among the members of a population. Organisms compete for available food, space, water, oxygen, sunlight, mates, and territory to raise young. In Africa, the ecology does not support enough vegetation for huge populations of elephants to survive and reproduce. Those elephants that are the best fit for the competition for resources will survive and reproduce. Those elephants that are less fit will not survive. Natural selection occurs when there is an interaction between organisms and their environment and among organisms living in a given environment. Natural selection limits the population growth.

2. Thomas Malthus, in 1798, warned that the human population was increasing so rapidly that it would soon be impossible to feed all of Earth's inhabitants. Darwin noted that food supply and other factors would hold populations in check. Human populations continue to grow at rapid rates in many countries of the world. Why do you think humans have avoided the pressures of selection indicated by Malthus and Darwin?
Answers may vary but the following points could be covered: The technology of food production and medicine and transportation have changed dramatically since the time of Malthus and Darwin. Selective breeding of improved crops and animals have been able to feed more peple than anticipated. Improved medical practices and medicines have kept people alive much longer than prior to the medical revolution. Humans have employed artificial selection in agriculture and human populations that have changed the conditions significantly from the period of time when either Malthus or Darwin lived.

3. How does selective breeding influence the process of evolution?
Selective breeding of any organism is a faster process for altering the frequency of genes in the population gene pool than through natural selection. Selective breeding chooses the desired traits of the plant or animal, breeds those organisms that possess the traits, and does not breed the organisms with the less desired traits. Within a few generations, plant and animal breeders are able to obtain offspring with the traits desired.

Copyright © by the Glencoe Division of Macmillan/McGraw-Hill School Publishing Company

EVOLUTION

Interpreting and Applying Concepts continued

Table 12–1 gives a partial amino acid sequence in hemoglobin proteins of humans, gorillas, and horses. Answer the following questions.

Table 12–1

HUMAN	Gly	Lys	Val	Asp	Glu	Val	Gly	Gly	Glu
GORILLA	Gly	Lys	Val	Asp	Glu	Val	Gly	Gly	Glu
HORSE	(Asp)	Lys	Val	Asp	(Glu)	Val	Gly	Gly	Glu
HUMAN	Lys	Leu	His	Val	Asp	Pro	Glu	Asp	Phe	Arg	Leu
GORILLA	Lys	Leu	His	Val	Asp	Pro	Glu	Phe	(Leu)	Leu
HORSE	Lys	Leu	His	Val	Asp	Pro	Glu	Asp	Phe	Arg	Leu

4. Circle those instances in Table 12–1 when the amino acids are not the same among the three organisms.

5. In this shortened sequence, how many times were there differences between the sequence of human and gorilla amino acids? __A difference occurred only once.__

6. How many times were there differences between the human and the horse? __Differences occurred 3 times.__

7. From this information and your knowledge of biology, which two organisms do you think evolved from the most recent common ancestor? __human and gorilla__

8. Give reasons for supporting or rejecting the following statement: Evolutionary relationships are stronger between organisms that have close biochemical similarities than between organisms that do not have close biochemical similarities.

   __Possible answer: Biochemical similarities between two organisms indicate a similarity in the DNA, which transcribes proteins that are similar between the organisms. If the DNA is similar, there is strong supporting evidence that there was a common ancestor from which the minor DNA modifications evolved.__

EVOLUTION

Interpreting and Applying Concepts continued

Look at Figure 12–1 and answer the question.

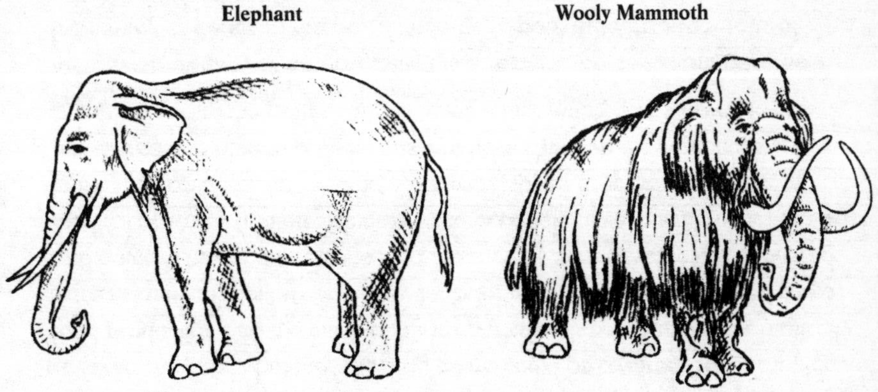

Elephant   Wooly Mammoth

Figure 12–1

9. According to biochemical tests, there is a close evolutionary link between the wooly mammoth, which roamed North America, Asia, and Europe thousands of years ago, and the modern elephant (Figure 12–1). What changes do you think took place for the elephant to have evolved from the ancestral mammoth?

   __Possible answer: The woolly mammoth flourished during the ice age and was well adapted to the climate. It had a heavy coat of fur and long tusks for protection against predators. Most likely, there was variation in the population of mammoths, with some having genes for heavier coats and some having genes that caused less heavy coats. Over time and changing environments, the selecting factors must have favored the individuals with genes for less heavy coats, because those individuals would have survived better in warmer climates. Over time, natural selection favored the mutations that coded for less fur. Today, elephants exist in naturally warm climates of Africa and Asia.__

CHAPTER 13 ASSESSMENT
ADAPTATION AND SPECIATION

Understanding Concepts

In the space at the left, write TRUE if the statement is true or FALSE if the statement is false.

true 1. The leaves on a tree are adapted for carrying on photosynthesis.

false 2. Any variation found in a plant or animal is considered an adaptation for survival of the individual.

true 3. Adaptations are the result of chance mutations that accumulate over a long period of time.

false 4. Horses and donkeys are considered to be in the same species.

true 5. A tetraploid (4n) plant is an example of a polyploid organism.

false 6. Homo sapiens are believed to have first appeared on Earth about 1.8 million years ago.

false 7. The idea that evolution occurs through slow and steady adaptive changes in populations is known as convergent evolution.

true 8. In the long process of evolution of the eye, a probable early step was the development of a group of light sensitive cells in simple organisms.

true 9. The poisonous venom of a snake is a physiological adaptation.

false 10. The oldest fossils of human ancestors were discovered in Europe.

Answer the following questions.

11. What is speciation? **Speciation is the process by which new species evolve from ancestral species.**

12. How does geographic isolation influence speciation? **If groups are isolated by geographic conditions, individuals of the two groups cannot interbreed. Different variations may become naturally selected and the two populations may gradually become distinct species.**

13. What are three adaptations common to all primates? **large brains, opposable thumbs, and rotating forelimbs**

14. Describe a behavioral adaptation of birds. **Possible answers: seasonal migration, nest building, parental care, territorial behavior, food seeking, incubation of eggs, social behavior, food gathering**

15. Compare gradualism and punctuated equilibrium. **Gradualism explains evolution as occurring through a process of slow and steady change in species. Punctuated equilibrium refers to a process of long periods of species stability followed by short periods of rapid evolution.**

Copyright © by the Glencoe Division of Macmillan/McGraw-Hill School Publishing Company

---

ADAPTATION AND SPECIATION

Interpreting and Applying Concepts

Answer the following questions in complete sentences.

1. Describe the evolution of finches on the Galapagos Islands, using the following terms in your answer: reproductive isolation, ancestral population, geographic isolation, adaptive radiation, gene pools.

Answers will vary but could include the following: It is thought that the portion of ancestral population of finches on the mainland of South America were blown away from the mainland resulting in geographic isolation on the many islands of the Galapagos. There were no competing species on the island for the finches. Variations that arose in the gene pools could develop into different species according to the selective pressures that favored those adaptations. The many beak forms evolved by adaptive radiation to the various sources of food found on the island. The geographically isolated finches could not interbreed and developed into different species that became reproductively isolated from each other.

2. Describe how a fire in a field could cause genetic drift in the populations of weeds that were left to reproduce.

Answers will vary but could be the following: When the gene pool changes drastically by chance, it is considered genetic drift. The fire in a field could have destroyed most of the weeds, leaving a small population that contained a gene pool that was not representative of the whole population of weeds. That small population would then produce offspring and the frequency of the alleles in the population would not be the same as the frequency of the alleles in the original population of weeds in the field.

3. If you were given a large box of primate skeletons, what are some factors that would help you classify them according to how each fit into a sequence of evolution of primates?

Answers will vary. Primate evolution shows a basic body plan but with modifications in total height, upright posture, and brain capacity by the size of skulls. The skeletons that demonstrated the tallest height, most erect posture, and the largest brain capacity would be the most recently evolved organisms. The trend would be from skeletons that were smaller, less erect, and with a small skull to the other extreme.

Copyright © 1994 by the Glencoe Division of Macmillan/McGraw-Hill Publishing Company

ADAPTATION AND SPECIATION

Interpreting and Applying Concepts continued

In the presumed evolution of the horse, paleontologists have been able to construct a reasonable picture of the probable sequence of ancestors to the modern horse. Study the time line in Figure 13–1 and the descriptions of early ancestors below it. In the space at the left of each description, write the letter that indicates about how many years ago each ancestor lived.

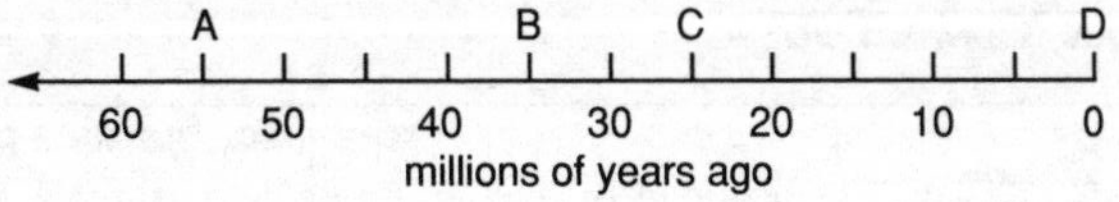

```
  A    B    C    D
|----|----|----|----→
60   50   40   30   20   10   0
       millions of years ago
```

Figure 13–1

B 4. This ancestor was much larger than a dog and both front and hind feet had three toes.

D 5. The modern horse is large and has one toe on each foot.

C 6. This ancestor was a grazing animal with three toes on each foot, the center toe being larger than the other two. The two smaller toes did not reach the ground.

A 7. This ancestor was about the size of a fox terrier. It had four toes on each front foot and three toes on each hind foot. It fed on trees and bushes.

Answer the following questions in complete sentences.

8. Describe what selective pressures favored the evolution of the horse to its present form.

Answers will vary, but could include that the variations that contributed to the increased size of the horse and to the modifications of the feet gave the horse an adaptive advantage to get away from predators, to run at great speeds without hurting the feet, and to seek a variety of foods in far-ranging habitats.

9. Figure 13–2 is a picture of the extinct dodo bird. Compare the dodo to any kind of bird that you know that can fly. Describe the adaptations found in the modern bird that were not evident in the dodo.

Dodo (extinct)

Figure 13–2

Answers will vary, but could include the facts that most modern birds have larger wings adapted to flight, hollow bones that allow them to be light for flight, smaller heads and beaks with beaks that are specific for the type of food eaten. Modern birds usually have thin legs.

Copyright © 1994 by the Glencoe Division of Macmillan/McGraw-Hill Publishing Company

CHAPTER 14 ❧ ASSESSMENT

CLASSIFICATION

Understanding Concepts

In the space at the left, write the letter of the word or phrase that best completes the statement or answers the question.

b   1. The scientific name *Homo sapiens* indicates that organisms given this name
 - a. belong to the species *Homo.*
 - b. belong to the genus *Homo.*
 - c. belong to the class *Homo.*
 - d. belong to the phylum *Homo.*

b   2. Which kind of evidence would help most to classify an organism?
 - a. the organism's habitat
 - b. the evolutionary relationships of the organism
 - c. the organism's eating habits
 - d. the function of the organism's body parts

d   3. Which is true of organisms classified in the same genus?
 - a. They are in the same order but in different families.
 - b. They belong to the same species but could be in different phyla.
 - c. They must be in the same kingdom but could be in different phyla.
 - d. They are in the same kingdom but may be of different species.

c   4. Which of the following is true?
 - a. The giant panda and the red panda are both related to raccoons.
 - b. The giant panda and the red panda are both related to bears.
 - c. The giant panda is related to bears and the red panda is related to raccoons.
 - d. The giant panda is related to raccoons and the red panda is related to bears.

Write the word or phrase that best completes the statement.

5. The science of classification is called   **taxonomy**   .

6. The most inclusive taxon is the   **kingdom**   .

7. In their classification, the cat and the dog diverge at the level of the   **family**   .

8. Carolus Linnaeus introduced a system of naming organisms called   **binomial nomenclature** .

9. Modern classification uses techniques of molecular biology, in particular the analysis of proteins and   **DNA**   .

In the space at the left, write TRUE if the statement is true or FALSE if the statement is false.

false   10. DNA can reveal close evolutionary relationships but cannot help to determine when two organisms began to diverge.

true   11. The evolutionary history of an organism is called its phylogeny.

false   12. The phylum that vertebrates belong to has two subphyla.

CLASSIFICATION

Interpreting and Applying Concepts

Both cats and dogs are grouped into the same kingdom, phylum, subphylum, class, and order. Based on this information, answer the following questions.

1. List some of the common structural features of cats and dogs.

   Answers will vary but may include common features such as a backbone, a tail, hair, mammary glands, fur, four limbs, sharp teeth.

2. Cats and dogs belong to different families, genera, and species. What can you infer from this? Give some examples to support your answer.

   Students may infer that dogs and cats must have some different structural features. As examples they may list relative size, shape of head and ears, climbing abilities, or retractability of claws.

Study Figure 14-1, which is a schematic drawing of the five-kingdom classification system, and answer the question.

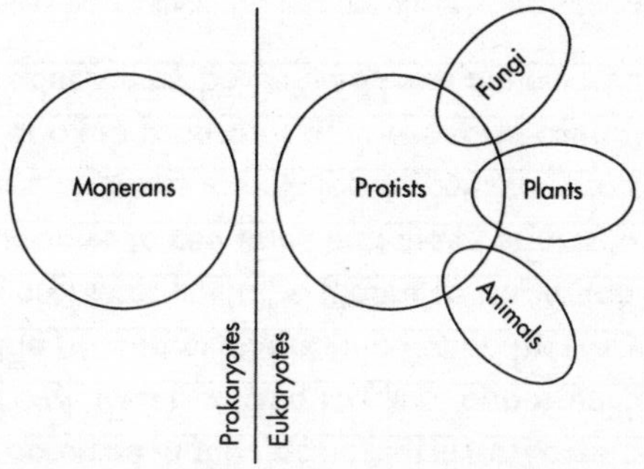

Figure 14-1

3. Explain how this drawing represents some of the problems with the five-kingdom classification system.

   Students may infer that the overlapping areas in the diagram shows that the classification system is not exact. They may explain that the system places unicellular eukaryotes in the protist kingdom and multicellular eukaryotes in the kingdoms of fungi, plants, or animals. However, there are some multicellular organisms in the fungi and animal kingdoms that are closely related to unicellular organisms; there are also some organisms, such as green algae, that may be unicellular or multicellular.

CLASSIFICATION

Interpreting and Applying Concepts   continued

Answer the following questions in complete sentences.

4. Devise a taxonomic key to distinguish among the following: apple, peach, pear, watermelon, orange, apricot, cherry, banana, grapefruit, nectarine.

Many different keys to distinguish the various fruits are possible.
Distinguishing features that might be used in constructing a key include
size, shape, color, whether or not the fruit has one large seed or many
smaller seeds, whether or not the fruit has a thin or thick peel.

5. Classification systems have changed over the years as scientists gain more expertise in methods of grouping organisms. One system classified fungi as plants. Describe why you think fungi are now classified in their own kingdom.

Answers may vary but could include: Fungi include mushrooms, shelf fungi
and yeast, all organisms that do not have roots, stems, and leaves or the
pigment chlorophyll. Fungi are heterotrophs; plants are autotrophs.

6. In your textbook, *Volvox*, *Spirogyra*, red algae, and *Ulva* are classified as protists, while in other textbooks they are classified as green plants. How do you think algae should be classified and why? **Answers will vary but should show a reasoned response.**
Some students may think that algae, because they lack roots, stems, and
leaves as specialized tissues, should be placed with the protists. Others will
suggest that because they have similar pigments and chloroplasts, they
should be classified as plants.

7. The red wolf, an inhabitant of the southeastern United States, was close to extinction in 1975 because of hunting and human encroachment on its habitat. Recently, some scientists have argued that the red wolf is a cross between the coyote and the gray wolf and so should not be placed on the endangered list. How could one demonstrate that the red wolf is a cross and not an endangered species?

Answers will vary but the evidence could be obtained by comparing the
DNA of the three animals. The DNA analysis would demonstrate the
possible hybrid condition of the animal. Other evidence might include close
examination of the fossil record of the three animals and their distribution.
In addition to this, the coyote and the gray wolf could be placed in a
mating situation to see if they produced a red wolf.

Copyright © by the Glencoe Division of Macmillan/McGraw-Hill School Publishing Company

---

CLASSIFICATION

Interpreting and Applying Concepts   continued

8. Maintaining species diversity is of great concern to environmentalists. Discuss reasons why more species are becoming endangered (you may cite specific examples) and suggest possible solutions for solving the problems.

This in an open-ended question that can lead to a variety of answers.
Students may mention that increasing populations make more demands on
the environment, emerging third-world countries rely on the use of the
environment for food and economic growth, and special interest groups
are willing to destroy the environment to protect jobs. Students may cite
such specific examples as rain forest destruction in South America, the
polarization over preserving the habitat of the spotted owl, thus reducing loggers' opportunities for work,
or the destruction of the African elephant population by ivory hunters.
Some solutions to the problems could be global conservation and
protection policies, subsidies to countries that make efforts to preserve
species, more emphasis on plans to develop regions for ecotourism,
economic investment and job programs in third-world countries where
medicines or other products may be derived from the environment.

9. A researcher reported the data below about the number of species in different orders of insects. Make a horizontal bar graph for the data.

Aptergota (wingless insects) 3500
Hemiptera (bugs) 60 000
Orthoptera (grasshoppers, locusts, crickets) 20 000
Coleoptera (beetles) 350 000
Lepidoptera (butterflies and moths) 120 000
Hymenoptera (wasps, ants, bees) 100 000
Diptera (two-winged flies) 120 000

Number of Species in Different Orders of Insects

Copyright © by the Glencoe Division of Macmillan/McGraw-Hill School Publishing Company

CHAPTER 15 ASSESSMENT
VIRUSES AND MICROORGANISMS

Understanding Concepts

In the space at the left, write the letter of the word or phrase that best completes the statement or answers the question.

c 1. All of the following are diseases caused by viruses EXCEPT
   a. AIDS.
   b. measles.
   c. *Pneumococcus* pneumonia.
   d. the flu.

c 2. Protozoans are classified on the basis of
   a. type of nutrition.
   b. type of environment.
   c. method of locomotion.
   d. method of reproduction.

b 3. Which of the following is a common disease-causing organism in Africa?
   a. paramecium b. Plasmodium c. radiolarians d. slime mold

c 4. Which of the following groups of organisms contains a eukaryotic organism that has been useful in recombinant DNA research?
   a. bacteria b. algae c. fungi d. protozoa

Write the word or phrase that best completes the statement.

5. Viruses are unique in that they can carry on but one life function, __reproduction__.

6. Some RNA viruses contain a specific enzyme, called __reverse transcriptase__, that allows them to make DNA from their RNA.

7. Methane-producing bacteria obtain energy by the process of __chemosynthesis__.

8. Bacteria that are important in the recycling of nutrients in organic wastes or dead organisms are called __saprophytes__.

9. Fungi are classified by the way in which they __produce spores during sexual reproduction__.

In the space at the left, write TRUE if the statement is true or FALSE if the statement is false.

false 10. Bacteria have flagella that have a structure and function identical to those found in protozoans.

false 11. The pigments in blue-green bacteria are found in chloroplasts, where photosynthesis takes place.

true 12. Brown algae are complex marine protists important as a commercial source of iodine.

false 13. The only mechanism used by fungi to obtain nutrients is saprophytic.

VIRUSES AND MICROORGANISMS

Interpreting and Applying Concepts

Answer the following questions in complete sentences.

1. A lab technician was culturing euglenoids. She covered the container with foil but cut an opening the size of a quarter in the side of the foil. She then placed a lamp in front of the container. The next day, she removed the foil to observe where the population of euglenoids was densest. Describe what you think she found and give a reason why.
__The euglenoids would have been found in greatest density near the quarter-sized opening in the foil. Sensitive to light, euglenoids have chloroplasts with which to carry on photosynthesis.__

2. Explain why it is difficult to classify the slime molds. __Slime molds have characteristics of both protozoans and fungi. A plasmodium engulfs organic matter by phago-cytosis, as does an amoeba. When food and moisture become scarce, a slime mold forms spore-producing structures similar to those in fungi. The spores are released and spread to other environments. Swarm cells that develop from germinated spores move by means of flagella, as do simple protozoans.__

3. Bacteria are found on surfaces everywhere--on hay, for example. If hay is placed in pond water containing a small population of paramecia, within one or two days the paramecium population will become fairly large. Explain how this happens.
__Answers may vary. Possible answers: When hay is placed in water, bacteria on the hay begin to decompose the hay rapidly. Because the hay provides a ready source of food, the bacteria reproduce rapidly. As the bacterial population increases, the bacteria form an abundant source of food for the paramecia. In turn, the paramecia reproduce rapidly.__

4. Dogs and cats suffer from viral infections, dogs from canine distemper and cats from feline leukemia, for example. Why do these diseases not spread to humans?
__The specific receptors on the membranes of the host cells are recognized by the receptors on the protein coats of the viruses. Cell membranes of humans do not contain the same receptor proteins as those found in dogs or cats. Thus, the virus would not recognize the human receptors and invade human cells.__ Viruses are frequently grown in developing chick embryos. Viruses produced in this way are used to make vaccines for diseases such as smallpox, influenza, and yellow fever. Why are chick embryos useful for the culture of viruses?

5. __Answers may vary. The following points should be made: Because viruses cannot live outside living organisms, they must be cultured in living tissue. Fertile eggs are readily available, and the virus can be contained inside the shell.__

VIRUSES AND MICROORGANISMS

Interpreting and Applying Concepts   continued

6. Viral DNA can be altered and new DNA can be spliced into the DNA of the virus. This is one technique used in recombinant DNA to transfer DNA into host cells. Explain why a virus would be a good way to transport new genes into host cells.

Answers may vary, but the following points should be covered: Because a virus can either enter a host cell itself or inject its DNA into a host cell, the virus would be a useful means of transporting genes from one place to another. Because viruses are specific for the host tissue, however, care would have to be taken when selecting the virus to be used in recombinant DNA techniques.

7. In what ways are fungi important to humans?

In addition to some types of mushrooms being used for food, many fungi have important uses in industry. For example, yeasts are used in baking and the brewing of beer. Fungi are also used to flavor cheeses.

8. In many tropical countries, malaria is a common and often fatal disease. Malaria is caused by a type of protozoan transmitted by the female *Anopheles* mosquito. Mosquitoes live and breed in stagnant water. Many have developed resistance to insecticides commonly used. Given these facts, how might the spread of malaria be prevented? **Answers may vary. Possible answers: developing new, stronger insecticides; removing water from ponds where mosquitoes breed; using screens or nets to protect humans from mosquitoes; altering the mosquitoes genetically through recombinant DNA technology; and developing methods of biological pest control (sterilizing males, introducing enemies of the mosquitoes)**

An investigation was performed to determine the rate of growth of a population of bacteria over a period of 16 hours. The data obtained by sampling the population every four hours is given in Table 15-1. Use the table to answer the questions that follow.

Table 15-1

Time In Hours	Number of Bacteria
0	1500
4	14 000
8	76 000
12	58 000
16	20 000

VIRUSES AND MICROORGANISMS

Interpreting and Applying Concepts   continued

9. According to Table 15-1, when was the rate of bacterial growth greatest?

between 4 and 8 hours

10. What explanation would you give for the change in population growth after 8 hours?

Answers may vary. Possible answers: The bacterial population may have decreased because of the depletion of food and oxygen supplies. The bacteria might have produced toxic wastes that caused the population to decline.

Read the following paragraph. Then answer the questions in complete sentences.

In April 1993, a number of people in Milwaukee, Wisconsin, became very sick as a result of contaminated drinking water. They contracted an influenza-like disease. Residents were told to boil water, and schools shut off drinking fountains. The disease, cryptosporidiosis, was caused by a parasite, *Cryptosporidium*, that is resistant to chlorine. The parasite produces egg cells that hatch only when they are in the digestive tract of a host. There, they produce a poison that causes diarrhea and vomiting. The disease is common in countries having poor sanitation and has occurred in places where people drink untreated or improperly treated water. Proper treatment of water supplies will remove the parasite.

11. How do you think the parasite might have gotten into the water supply in Milwaukee?

Answers may vary. In countries with poor sanitation, the parasite is probably transmitted by human wastes that get into the water supply. Although the means of transmission in Milwaukee has not yet been fully determined, human wastes might have been discharged into Lake Michigan, which provides the water supply for Milwaukee. Also, if the treatment system for drinking water were not working properly, parasites would not be killed.

12. Why might it be difficult to inspect water samples in order to determine the presence of *Cryptosporidium*?

The parasite is in a dormant state until it reaches the digestive tract of a host.

PLANT ADAPTATIONS

Interpreting and Applying Concepts

All living things possess characteristics that form the basis of their classification. The table below shows certain characteristics of green algae, nonvascular plants, and vascular plants. Based on the data in Table 16–1, answer the questions that follow.

Table 16–1

Characteristics of Green Algae, Nonvascular Plants, and Vascular Plants

Organism	Characteristics
Green Algae	1. Manufacture food through photosynthesis 2. Have cell walls containing cellulose 3. Make starch 4. Limited to watery or moist environment
Nonvascular Plant	1–4 above plus: 5. Possesses a cuticle that inhibits water loss 6. Takes in carbon dioxide from air
Vascular Plant	1–3 and 5–6 above, plus: 7. Possesses structures that provide support 8. Possesses true roots, stems, and leaves 9. Usually grows very tall 10. Can live in a wide variety of environments

1. What data in Table 16–1 would support the hypothesis that plants evolved from green algae? **Although plants are more complex than green algae, they share three basic characteristics of green algae; Characteristics 1–3.**

2. If you discovered a new plant species in a desert, what inference could you make concerning its probable classification? Why? **The organism is probably a vascular plant because it inhabits a dry environment. Nonvascular plants are generally limited to moist environments.**

3. Suppose you have a specimen of a single cell from an organism that you have not observed. You examine the cell under a microscope. You discover that the cell has a rigid cell wall and chlorophyll. What conclusions could you reach, or not reach, concerning the classification of the organism from which the cell came? Why? **Since the cells of green algae as well as those of plants possess walls and chlorophyll, you could not classify the organism from which it came as either a green alga or a plant.**

Copyright © by the Glencoe Division of Macmillan/McGraw-Hill School Publishing Company

---

CHAPTER 16 ✿ ASSESSMENT
PLANT ADAPTATIONS

Understanding Concepts

In the space at the left, write the letter of the word or phrase that best completes the statement or answers the question.

b 1. All modern plants are probably descendants of
  a. fungi. b. green algae. c. club mosses. d. horsetails.

d 2. Protists include
  a. ferns. b. gymnosperms. c. angiosperms. d. green algae.

c 3. An adaptation of land plants that inhibits water loss is the
  a. cell wall. b. chloroplast. c. cuticle. d. vascular tissue.

d 4. During photosynthesis, plants
  a. take in oxygen and carbon dioxide.
  b. take in oxygen and give off carbon dioxide.
  c. give off oxygen and carbon dioxide.
  d. take in carbon dioxide and give off oxygen.

b 5. Rhizoids are structures analogous to
  a. stems. b. roots. c. leaves. d. chloroplasts.

b 6. The sporophyte generation of plants consists of
  a. *n* cells. b. 2*n* cells. c. 3*n* cells. d. 4*n* cells.

a 7. Sporangia are adaptive characteristics of
  a. ferns. b. green algae. c. gymnosperms. d. angiosperms.

a 8. Seeds enclosed in a fruit are adaptations of
  a. angiosperms. c. nonvascular plants.
  b. gymnosperms. d. green algae.

d 9. The plants best adapted to survival in harsh, dry climates are
  a. green algae. b. ferns. c. liverworts. d. gymnosperms.

In the space at the left, write TRUE if the statement is true or FALSE if the statement is false.

false 10. In flowering plants, the male reproductive organs are the stigmas.

true 11. Tree bark is an example of a plant adaptation to reduction of water loss.

false 12. Nonvascular plants are well-adapted for growth in dry environments.

false 13. Gymnosperms have reproductive structures in needle-shaped leaves.

true 14. In a moss, the cells of the gametophyte generations are haploid.

PLANT ADAPTATIONS

Interpreting and Applying Concepts continued

4. A few days later, you obtain a larger sample of the organism. You find the specimen is covered with a cuticle. What can you now infer about the classification of the organism? Why? **The organism is either a nonvascular or vascular plant since only plant tissues contain a cuticle.**

5. You really want to pin down the classification of this mysterious organism. You go out into the field in search of the whole plant. Unfortunately all you find is a clump of soil in which you discover the organism's root. This clue allows you to classify the organism into a single group; what group and why? **The organism is a vascular plant since only vascular plants possess roots.**

While on a field trip, you collect an assortment of plant parts. You label each part with a code number. The parts and their numbers are shown in Figure 16–1. When you get back to your lab, you complete the classification table, Table 16–2. After completing the table, answer the following questions.

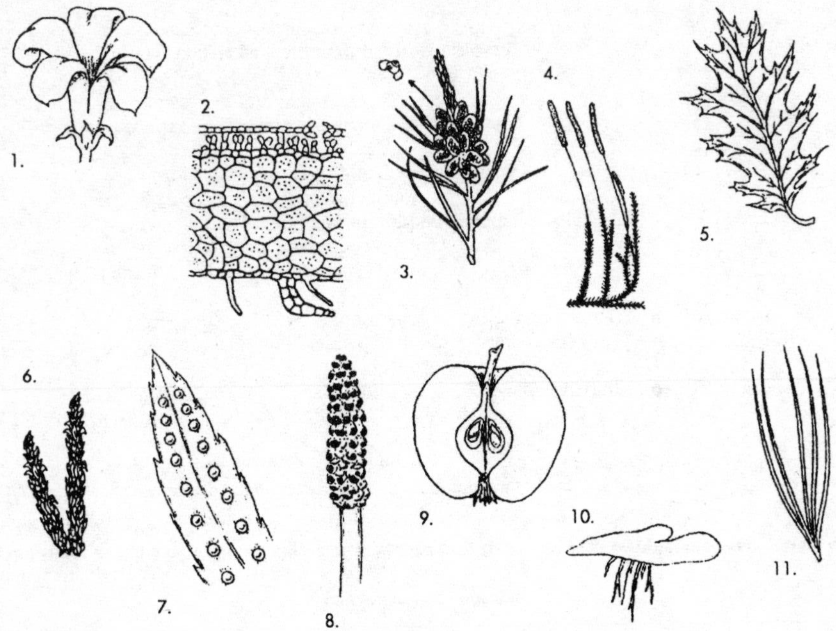

Figure 16–1

PLANT ADAPTATIONS

Interpreting and Applying Concepts continued

Table 16–2

Classification Table		
Organism	Code Number	Description/Identification
Moss	6	gametophyte of a moss
Liverwort	2	thallus of a liverwort
Club moss	4	sporophyte generation of the club moss
Horsetail	8	sporophyte of a horsetail
Fern	7, 10	(7) underside of a fern frond (10) gametophyte of fern
Gymnosperm	3,11	(3) cone of gymnosperm (11) leaf of gymnosperm
Angiosperm	1, 5, 9	(1) angiosperm flower (5) leaf of angiosperm (9) angiosperm fruit

1. Which of the collected structures might contain pollen? **Gymnosperms produce pollen in cones (3) and angiosperms produce pollen in flowers (1).**

2. Among the seed-bearing plants, infer from the drawings which have leaves that are best adapted to a windy, relatively dry environment. Why? **The leaves of the gymnosperms possess a relatively small surface area that helps to inhibit water loss, which a windy, relatively dry climate promotes.**

3. Which structure is an adaptation that protects seeds and allows its parent plants to be classified as the most successful plants on Earth? **the fruit (9) of the angiosperm.**

152

ANIMAL ADAPTATIONS

Understanding Concepts

In the space at the left, write **TRUE** if the statement is true or **FALSE** if the statement is false.

true	1.	Sponges are classified according to the structure and composition of spicules.
false	2.	In sponges, reproduction from body fragments results in genetic variability.
true	3.	Tissues in cnidarians are evidence of an evolutionary advance over sponges.
false	4.	In general, bilateral symmetry is typical of the most primitive animals.
false	5.	Flatworms are the first animals in the evolutionary tree to possess two body openings.
true	6.	Flatworms reproduce by cross-fertilization, which ensures genetic variability.
false	7.	Each segment in a segmented worm contains different structures.
false	8.	Segmented worms possess a primitive brain but no nerve cord.
true	9.	Some mollusks do not have shells.
true	10.	The appendages of an arthropod allow for a variety of movements.
true	11.	Internal fertilization is an adaptation of land-dwelling animals.

Answer the following questions.

12. In what way is the decentralization of an echinoderm's nervous system an adaptive advantage? **It allows animals with radial symmetry to respond to stimuli coming from any direction.**

13. What led taxonomists to separate the hemichordates and chordates into two different phyla? **Although both possess nerve cords, the structures are not homologous.**

14. What is the major characteristic that distinguishes organisms in the subphyla that contain tunicates and lancelets from vertebrates? **Vertebrates possess a backbone, which is not possessed by animals in the other subphyla.**

15. What distinguishes the two major groups of vertebrates? **the presence or absence of legs; Fish do not have legs. All other vertebrate classes contain at least some animals that have legs.**

16. What is the main difference between the reproductive process of amphibians and those of reptiles, birds, and mammals? **Fertilization in amphibians occurs externally while that in reptiles, birds, and mammals occurs internally.**

ANIMAL ADAPTATIONS

Interpreting and Applying Concepts

Figure 17-1 presents data for the eight vertebrate classes. Based on the chart, answer the following questions.

Figure 17-1

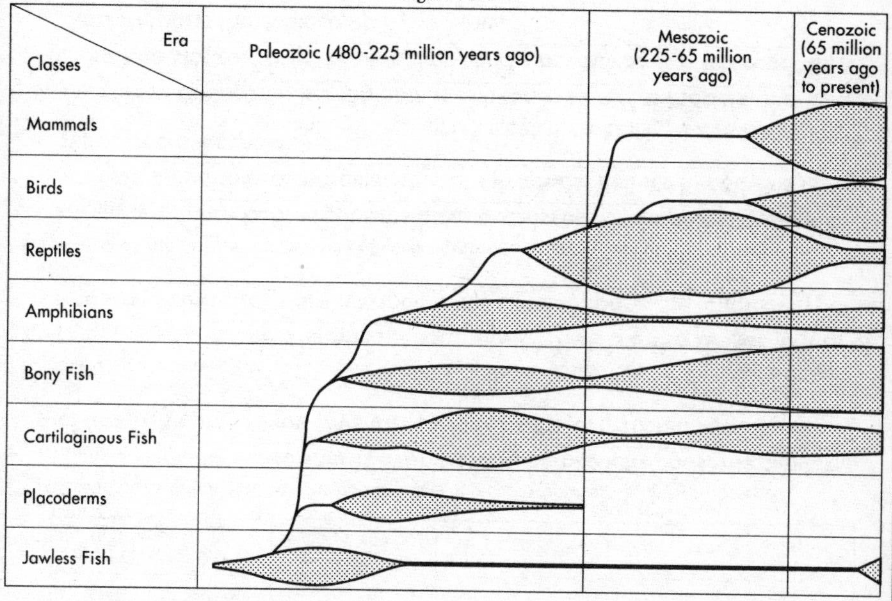

Classes	Era — Paleozoic (480-225 million years ago)	Mesozoic (225-65 million years ago)	Cenozoic (65 million years ago to present)
Mammals			
Birds			
Reptiles			
Amphibians			
Bony Fish			
Cartilaginous Fish			
Placoderms			
Jawless Fish			

1. Suggest an appropriate title for the chart. **Answers may vary. A possible title is "Changing Populations of Vertebrate Classes over the last 500 Million Years."**

2. A distinguishing characteristic of birds and mammals is that they are endothermic. Express a hypothesis and the reasoning behind it, to support the possibility that extinct members of another class were endothermic. **A reasonable hypothesis might be that some dinosaurs or other early reptiles were endothermic. This hypothesis would find theoretical support from the fact that birds and mammals evolved from early reptiles, some of which were dinosaurs.**

153

ANIMAL ADAPTATIONS

Interpreting and Applying Concepts   continued

Table 17–1

Characteristics of Coelacanths, Lungfish, and Tetrapods		
Coelacanth	Lungfish	Tetrapod
gills, degenerate lungs	gills and air-breathing lunglike structures	air-breathing lungs in all adults
limbs are leglike	limbs are leglike	all classes have members that have legs
hemoglobin part similar to tadpole's	—	—
—	mitochondrial DNA more similar to frog's than coelacanth's mitochondrial DNA	—
—	ear like shark's	—
control of eyes more like that of tetrapods	control of eyes less like that of tetrapods	—

7. Based on the family tree and the datum regarding ear structure, which hypothesis is supported? Why? **Since sharks evolved long before tetrapods, the similarity of lungfish ears to those of sharks supports the coelacanth hypothesis.**

8. Based on the family tree and the data regarding mitochondrial DNA, which hypothesis is supported? Why? **Since lungfish mitochondrial DNA is more similar to that of frog's and since frogs are tetrapods, this data supports the lungfish hypothesis.**

9. Why does the respiratory anatomy of coelacanths and lungfish support or not support either hypothesis? **Since both organisms have possessed, or do possess, lunglike structures, which are characteristics of tetrapods, neither hypothesis is supported over the other.**

10. Based on the family tree and the data regarding hemoglobin, which hypothesis is supported? Why? **Since coelacanth hemoglobin is similar to that of a tadpole, the immature stage of frog development, and since a frog is a tetrapod, this data supports the coelacanth hypothesis.**

11. Where on the family tree do you believe the coelacanth and lungfish should be? Explain. **Answers will vary but should be logically supported. For example, students may rely more strongly on chemical rather than structural evidence.**

Copyright © by the Glencoe Division of Macmillan/McGraw-Hill School Publishing Company

---

ANIMAL ADAPTATIONS

Interpreting and Applying Concepts   continued

3. The chart implies that amphibians evolved after fish but before reptiles. What characteristic of frogs supports this implication? How does it do so? **In their tadpole stage, frogs are water-dwellers and like fish, use gills to breathe. In their adult stage, frogs are land-dwellers and like reptiles, use lungs to breathe.**

4. How can the data in the chart be interpreted to reveal which class of reptiles is extinct today? **The placoderms are extinct. Their timeline ends 225 million years ago.**

5. Interpret the chart to determine which class gave rise to two other classes. What are the classes and which gave rise to which? **The reptiles gave rise to mammals and birds.**

6. What era could be considered the age of reptiles? Use the given data to support your answer. **The age of reptiles would be the era during which the population of this class was greater than that of any other class. That era was the Mesozoic.**

Scientists are involved in a controversy involving the positioning of two fish, the coelacanth and the lungfish, on the vertebrate family tree. Scientists have proposed two conflicting hypotheses about these fish. One suggests that the coelacanths are more closely related to tetrapods than to lungfish. The other suggests that the opposite is true. Figure 17–2 shows the vertebrate family tree with two branches unoccupied. Table 17–1 reveals data concerning the coelacanth, lungfish, and tetrapod. Based on the figure and the table, answer the questions that follow.

Figure 17–2

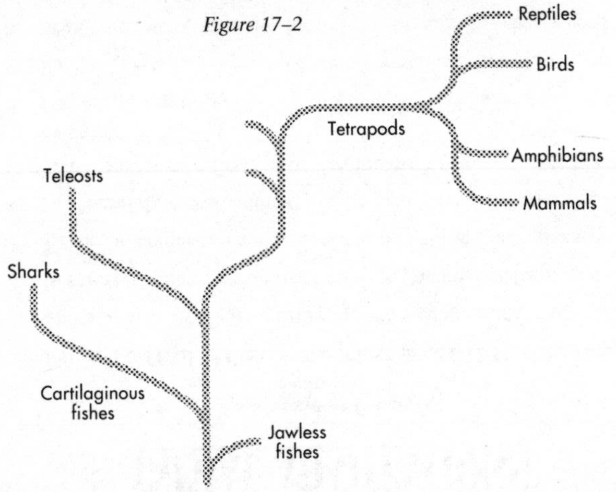

Copyright © by the Glencoe Division of Macmillan/McGraw-Hill School Publishing Company

154

CHAPTER 18 ⚬ ASSESSMENT

REPRODUCTION

Understanding Concepts

In the space at the left, write the letter of the word or phrase that best completes the statement or answers the question.

b   1. Bulbs are food storage organs that aid a plant in
    a. spreading out over a large area.   c. supplying food for animals and people.
    b. surviving over winter.   d. resisting infection.

a   2. Vegetative reproduction in a creeping stem differs from the growth of an ordinary stem because after a creeper is rooted, it
    a. separates from the parent plant and grows independently.   c. dies off.
    b. grows faster.   d. produces seeds.

c   3. Fragmentation is a kind of asexual reproduction for starfish because they can
    a. break open a clam.   c. regenerate all the missing parts of their body from a single small part.
    b. escape by leaving a limb behind.   d. break into many parts.

d   4. A gardener grows a new plant from a cutting by means of
    a. budding.   b. corms.   c. tubers.   d. regeneration.

b   5. An animal produced by parthenogenesis develops from
    a. a fertilized egg.   c. a regenerated part.
    b. an unfertilized egg.   d. diploid spore.

In the space at the left, write the letter of the phrase from Column B that best matches the term in Column A.

	Column A		Column B
c	6. budding	a.	transfer of pollen from anther to stigma
a	7. pollination	b.	genetic material transferred by cell-to-cell contact
f	8. estrus	c.	method of reproduction in *Hydra*
g	9. fragmentation	d.	individual in which both kinds of sex organs develop
j	10. ovulation	e.	energy source in some new, developing plants
i	11. gametes	f.	period of readiness for mating
b	12. conjugation	g.	method of reproduction in planarian
h	13. ovules	h.	become seeds after fertilization
e	14. endosperm	i.	sex cells
d	15. hermaphrodite	j.	release of an egg from a follicle

REPRODUCTION

Interpreting and Applying Concepts

Animals that reproduce by means of external fertilization have various adaptations to ensure fertilization of the eggs. The stickleback, a small fish, has been studied to find out what factors affect its reproductive success. Stickleback males build a nest and try to induce females to lay their eggs in the nest. Two factors seem to influence a female's choice of a male: the male's coloration and the location of the site selected by the male to build his nest. Study the graphs in Figure 18–1 and answer the following questions in complete sentences.

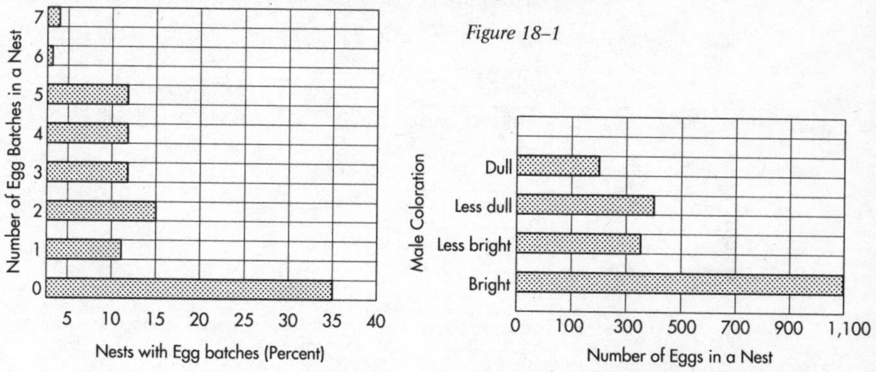

Figure 18–1

1. What percentage of males' nests received no batches of eggs?   <u>close to 35 percent</u>

2. What percentage of the nests received 4 or more egg batches?   <u>27 percent</u>

3. What could you conclude from the graph about the success of some males in attracting females?   <u>Answers may include that some males are highly successful in attracting females with eggs and others are not successful at all.</u>

4. What does the male coloration graph seem to show about the reproductive success of brightly colored males over males with less bright coloration? <u>The brightly colored males were 3 times as successful as less bright males.</u>

5. In the male coloration graph, less dull sticklebacks attracted females to lay more eggs in their nests than were laid in the nests of less bright sticklebacks. What factors besides color may be affecting the data? <u>Answers may include that all females may not lay the same number of eggs. The less dull males may have attracted more females that laid a larger number of eggs than the females attracted by the less bright males. The nests of the less dull males might have been better located than the nests of the less bright males.</u>

REPRODUCTION

Interpreting and Applying Concepts   continued

6. There are other factors that affect the mating behavior of sticklebacks. Males often eat the eggs fertilized by their competitors, and females often eat eggs laid by other females. How would these behaviors affect the data on the graphs? **A male stickleback may mate successfully, but this would not be recorded if other sticklebacks ate the eggs in his nest.**

In Australia, the aquatic snail *Potamopyrgus antipodarum* was found to have individuals that reproduced asexually and others that reproduced sexually. These snails are commonly infected with a parasitic worm. To find out if a snail's method of reproduction affects the degree to which it becomes infected with the worms, snails and worms were collected from two lakes, A and B, located 3050 m apart. Snail populations from both lakes had individuals that reproduced sexually and others that reproduced asexually. Use the diagrams in Figure 18–2 to answer the questions.

Figure 18–2

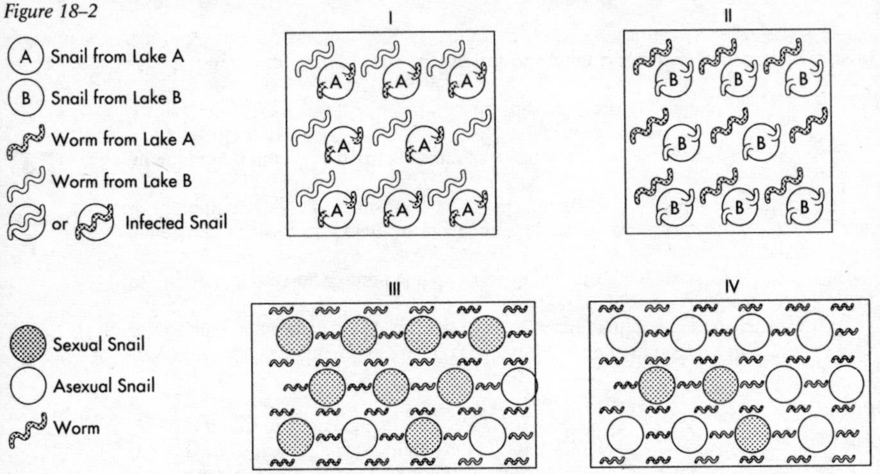

(A) Snail from Lake A
(B) Snail from Lake B
    Worm from Lake A
    Worm from Lake B
    or   Infected Snail

I   II   III   IV

Sexual Snail
Asexual Snail
Worm

7. Snails from Lake A were placed in container I with worms from both lakes. What happened to the snails? **The snails from Lake A became infected by the worms from Lake A but not by the worms from Lake B.**

8. When snails from Lake B were placed in container II with worms from both lakes, what happened to the snails? **The snails from Lake B became infected by the worms from Lake B but not from Lake A.**

9. A number of snails that reproduce sexually were placed in container III with an equal number of snails that reproduce asexually and a large number of worms. What happened to the asexual snails after a prolonged period of time? **The asexual snails became outnumbered by the sexual snails.**

Copyright © by the Glencoe Division of Macmillan/McGraw-Hill School Publishing Company

REPRODUCTION

Interpreting and Applying Concepts   continued

10. In container IV, equal numbers of sexual snails and asexual snails were placed with a small number of worms. What happened to the asexual snails after a prolonged period of time? **The asexual snails eventually outnumbered the sexual snails.**

11. Hypothesize why these changes in the kind of reproductive behavior were selected for. **When there are many parasites to infect the snails, the genetic diversity provided by the sexual reproduction makes them less vulnerable to parasites. When there are fewer parasites to contend with, the faster method of asexual reproduction takes over.**

12. Why might clones be easy prey for parasites? **Parasites usually coevolve with their hosts. It is easier for the parasite when the host produces only clones. The parasite finds it easier to keep up with clones than with a diverse population.**

Copyright © by the Glencoe Division of Macmillan/McGraw-Hill School Publishing Company

CHAPTER 19 ASSESSMENT
DEVELOPMENT

Understanding Concepts

In the space at the left, write the letter of the word or phrase that best completes the statement or answers the question.

b 1. Embryos of both plants and animals are similar in
    a. the timing of early cell division and growth of the embryo.
    b. having a food source for the embryo.
    c. development that results from different rates and patterns of cell division.
    d. development that results from cell movement.

c 2. Morphogenesis occurs when the
    a. egg is fertilized.
    b. zygote begins to divide.
    c. outer cells move inward and the first evidence of body systems appears.
    d. unfertilized egg is produced.

a 3. During the third trimester of human gestation, the fetus increases
    a. its mass.
    b. the number of body systems.
    c. the complexity of its body.
    d. its endoderm tissue.

c 4. Unlike animals, many plants can develop new structures throughout their lifetime because
    a. they never stop being embryos.
    b. if they stop developing, they die.
    c. they have regions of cell division, called meristems, in many parts of the plant.
    d. they make sugars that provide energy.

Write the word or phrase that best completes the statement.

5. The group of cells produced as a result of cleavage is called the **blastula**.

6. The **neural tube** is the beginning of the nervous system in the developing embryo.

7. A seed develops into a new plant during **germination**, which occurs if conditions are favorable for growth.

8. The **cotyledon** becomes the main food source for the developing plant embryo after the endosperm is depleted.

9. When one part of an embryo influences the development of another part, **embryonic induction** occurs.

10. The stage of development represented by a tadpole is called a(n) **larva**.

11. The **placenta** is tissue formed from both the mother's uterus and the embryo itself, and is the site of nutrient and waste exchange.

Copyright © by the Glencoe Division of Macmillan/McGraw-Hill School Publishing Company

---

DEVELOPMENT

Interpreting and Applying Concepts

Not all frogs develop in the usual way from egg to tadpole to four-legged adult. In fact, at least fourteen of the twenty-four known families of frogs show variations in the way they develop. These variations fall into three categories: (1) direct development, (2) transport of eggs or tadpoles by one of their parents, (3) development in foam nests. Direct development means developing from eggs to four-legged froglets without a tadpole stage. Frogs, male or female, that transport eggs or tadpoles may do so on their back, in their vocal sacs, or even inside their stomach! For those eggs covered by foam when they are laid, the outer layer of the foam hardens, providing greater protection. Study the diagram in Figure 19-1 which shows data for some families of frogs. Then answer the following questions.

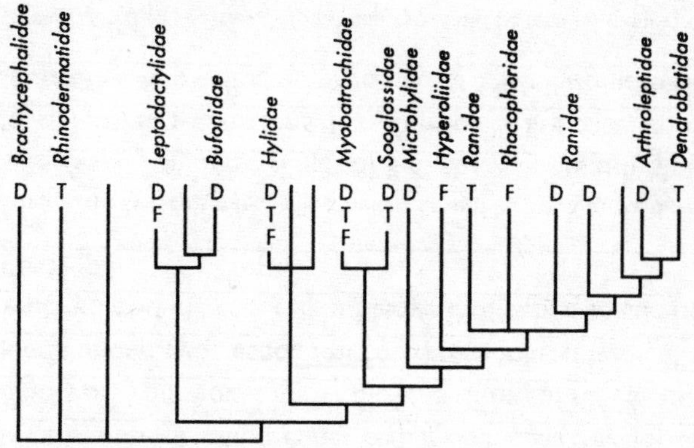

Figure 19-1

D   Direct development
T   Transported by a parent during development
F   Development in a foam nest
Lines show relationships

1. How many of the families of frogs shown in the diagram have species that develop directly from eggs to froglets without the tadpole stage? **Ten families have species that develop directly from eggs to froglets.**

2. What selective advantage might there be in direct development for frogs? **When the frogs bypass the tadpole stage during development, they probably increase their chance of survival. The tadpole stage is vulnerable to all kinds of predators.**

3. How many of the frog families shown have more than one unusual strategy for development? **Four of the frog families have either two or three unusual strategies for development.**

Copyright © by the Glencoe Division of Macmillan/McGraw-Hill School Publishing Company

Interpreting and Applying Concepts   continued

4. What factors may select for one type of development over another?
Factors in the environment such as moisture or lack of it, or the kinds of development strategies may affect the kinds of development strategies used.

5. When would protective foam around the eggs provide a selective advantage?
When the environment is warm and dry, the foam would keep the eggs from drying out. The hardened foam layer may also fool predators into thinking the eggs are not edible.

6. What advantage is there to one of the parents carrying the eggs or tadpoles on or in its body?
The eggs or tadpoles in or on a parent's body are protected from predators and may also be kept from drying out.

7. Use Figure 19–1 to tell whether or not family relationships affect the kinds of development strategies a family engages in. **Answers may include that although some effect of family relationships may be seen in the way development strategies are practiced, it is not clearly defined. It would seem that all the strategies are options for the different families and that the type of environment may play a more important role in determining which one is used than do family relationships.**

Read the following paragraph and answer the questions in complete sentences.

Scientists have tried to understand how an animal with many complex systems develops from a single-celled zygote. In the late nineteenth century, Wilhelm Roux, a German embryologist, studied the development of fertilized frog eggs. He investigated what would happen if he killed one cell of a two-celled embryo with a hot needle. Roux found that the live cell developed as half an embryo. Roux concluded that the abnormal development occurred because half of the material needed to determine a complete embryo was missing after he killed one cell. Hans Driesch performed a somewhat similar experiment with the two-celled stage of sea-urchin embryos. Instead of killing one of the cells, Driesch separated the two cells and observed what happened. Both cells continued to develop and to form two complete embryos, smaller than usual, but otherwise normal.

8. What was the purpose of the experiments carried out by Roux and Driesch?
The purpose of the experiments was to determine if development of a complete organism could take place from only part of an embryo.

Interpreting and Applying Concepts   continued

9. Most of the scientists in the eighteenth and early nineteenth centuries believed that all the embryo parts were complete (or completely determined) at fertilization rather than developing later. What do you think was the impact of each of the two experiments on this theory? **Answers will vary. Students might conclude that Roux's experiment seemed to uphold the theory because a half embryo developed from half of a two-celled embryo. They might conclude that the Driesch experiment did not uphold the theory because each half of the two-celled embryo developed into a complete embryo.**

10. Perhaps you concluded that the difference in the results of the two experiments occurred because Roux used frog eggs and Driesch used sea-urchin eggs. How could you prove that the kinds of eggs used were not the cause of the different results? **Answers may include that you could take a two-celled frog embryo and separate the cells without killing either cell. If both cells developed into small frog embryos, you would know that the different results were not due to the kinds of embryos used. You could also take a two-celled sea-urchin embryo and kill one cell but leave it attached to see how that cell would develop.**

11. Is it fair to say that at the two-cell stage, adjustments can still be made to the developmental process? Explain. **Yes, it can be said that at that early stage, adjustments can still be made, at least in sea urchins. If this were not the case, you would not be able to separate a two-celled embryo and produce two normal embryos.**

12. Beatrice Mintz investigated to see if one mouse embryo could be formed from two embryos. She removed the membrane around two eight-celled embryos and allowed the two embryos to fuse. After fusion, she placed the new embryo in the uterus of a mouse. A normal mouse was born. What does this investigation prove about early development in mice? **It proves that mouse embryos at the 8-cell stage have not yet begun to differentiate. In this case, the two embryos were altered to develop as just one embryo.**

NUTRITION AND DIGESTION

Interpreting and Applying Concepts

Iron is one of the minerals important to the health of the body. Although many foods contain iron, the body does not absorb the iron from plants as well as it does the iron from animal products. Study Table 20–1, which lists both plant and animal sources of iron. The table shows how much iron a 90-gram portion of each food contains and how much of this iron can be absorbed by the body. Answer the following questions in complete sentences.

Table 20–1 **SOURCES OF IRON**

	Iron Content (mg)	Iron Absorbed (mg)	% of Iron Absorbed
Iceberg lettuce	0.5	0.02	4.0
Spinach (cooked)	2.0	0.04	2.0
Black beans (cooked)	4.3	0.07	1.6
Soybeans (cooked)	5.0	0.35	7.0
Chicken (roasted, no skin)	1.2	0.22	18.3
Ground beef (broiled)	3.0	0.60	20.0
Sirloin steak (broiled)	5.4	1.08	20.0
Calf liver (fried)	12.0	1.80	15.0

1. Which food from plants is the best source of iron? Explain. **Cooked soybeans are the best source of iron because they have the highest iron content with the highest percent of iron absorbed.**

2. Explain why lettuce is not a good source of iron even though the percent of iron absorbed is higher than that of some of the other plants. **Even though the percent of iron absorbed is relatively high for lettuce, the iron content is very small. A person would have to eat a very large amount of lettuce to get the same amount of iron as in a serving of , for example, black beans.**

3. Which meat source would you choose for its iron content and ability to be absorbed by the body? **Fried calf liver would be the best choice for its iron content and ability to be absorbed by the body.**

4. Use the information about iron from plant and animal sources to explain why many people from poorer regions of the world suffer from an iron deficiency. **In the poorer regions of the world, people's diets include a greater proportion of food from plants, and often very little meat.**

5. Vitamin C helps and tea hinders the absorption of iron from plants. Based on this information, what recommendation might you make to a vegetarian friend? **Answers may include advising the friend to drink orange juice instead of tea when eating vegetables so that more of the iron will be absorbed.**

Copyright © by the Glencoe Division of Macmillan/McGraw-Hill School Publishing Company

---

CHAPTER 20   ASSESSMENT

NUTRITION AND DIGESTION

Understanding Concepts

In the space at the left, write the letter of the word or phrase that best completes the statement or answers the question.

b   1. Earthworms lack teeth but are able to grind their food into pieces in the
   a. intestine.   b. gizzard.   c. stomach.   d. mouth.

a   2. An animal that digests its food in a gastrovascular cavity provides a simple example of
   a. extracellular digestion.
   b. intracellular digestion.
   c. digestion without enzymes.
   d. a digestive tract with two openings.

c   3. The end product of protein digestion is
   a. bile.   b. maltose.   c. amino acids.   d. monosaccharides.

d   4. One end product of the digestion of fats is
   a. lipase.   b. cholesterol.   c. glucose.   d. glycerol.

d   5. Examples of heterotrophs that carry out intracellular digestion are
   a. *Rhizopus* and mushrooms.
   b. Venus's-flytrap and the pitcher plant.
   c. *Hydra* and the earthworm.
   d. *Amoeba* and *Paramecium*.

Write the word or phrase that best completes the statement.

6. The digestion of starch begins in the **mouth** and continues in the **small intestine**.

7. Protein digestion begins in the **stomach** and continues in the **small intestine**.

8. Fats are digested in the **small intestine**.

9. The surface area of the small intestine is increased by **villi** in its lining.

10. The **liver** converts excess glucose in the blood to glycogen.

In the space at the left, write TRUE if the statement is true or FALSE if the statement is false.

false   11. Plants do not require the same kinds of organic substances that heterotrophs need in order to grow.

true   12. Heterotrophs obtain food by eating other organisms.

true   13. In filter-feeding mollusks, both intracellular and extracellular digestion occur in the stomach and intestines.

false   14. After digestion, excess amino acids are converted to proteins.

Copyright © by the Glencoe Division of Macmillan/McGraw-Hill School Publishing Company

NUTRITION AND DIGESTION

Interpreting and Applying Concepts continued

Athletes at an advanced level of physical fitness have different nutritional needs from those at an intermediate level. The athletes at each level have a set Calorie requirement based on their physical activity. They keep track of the food units to be consumed daily, based on the food pyramids in Table 20-2. Answer the following questions in complete sentences.

Table 20-2 **NUTRITIONAL NEEDS FOR ATHLETES**

Food Category (1 Serving = 1 Unit)	Units/Advanced Level Female 2200 Cal.	Male 2800 Cal.	Units/Intermediate Level Female 1900 Cal.	Male 2500 Cal.
Meat (1 oz)	6	7	5	7
Dairy (1 cup)	3	3	2	2
Fat (1 tsp)	7–8	10–11	6–7	8–9
Fruit (1)	3	4	3	4
Vegetables (leafy – 1 cup; other – 1/2 cup)	4–4.5	4.5–5.5	4–4.5	4.5–5.5
Grain (1 slice bread; 1 cup dry cereal; 1/2 cup cooked cereal, pasta, rice)	9	13	8	12

6. Which food units are reduced for male athletes from the advanced to the intermediate level? **At the intermediate level, the male athletes eat 1 less dairy and grain units and 2 less fat units than males at the advanced level.**

7. How many more additional overall food units does a female athlete at the advanced level have to consume compared to a female athlete at the intermediate level? **At the advanced level she would have to consume 4 more units.**

8. Some athletes also work out at the beginner level. Suppose the number of Calories needed for the beginner level is decreased by the same amount that the number of Calories at the intermediate level is decreased from the advanced level. How many Calories will the beginner female and male athletes consume? **The female athletes will consume 1600 Calories, 2200 Calories and the male athletes, 2200 Calories.**

9. Female athletes at the beginner level will eat 4 food units less, equally divided between the fat and grain categories, than female athletes at the intermediate level. How many units of fats and grains will beginner female athletes consume? **They will consume 4–5 fat units and 6 grain units.**

10. If the male athletes at the beginner level need the same number of food units as the female athletes at the advanced level, how many fat and grain units will beginner male athletes consume? **The male athletes at the beginner level will consume 7–8 fat units and 9 grain units.**

Copyright © by the Glencoe Division of Macmillan/McGraw-Hill School Publishing Company

---

NUTRITION AND DIGESTION

Interpreting and Applying Concepts continued

The tally for the food units consumed by Jennifer and Jeff during three main meals are listed on the chart below.

	Meat	Dairy	Fat	Fruit	Vegetable	Grain
Jennifer	6	1	7	2	4.5	7
Jeff	7	2	10	2	5.5	9

11. If both Jennifer and Jeff are at the advanced level, what additional food units should they choose for snacks? **Jennifer should choose 2 dairy units, 1 fruit, and 2 grains. Jeff should choose 1 dairy unit, 2 fruits, and 4 grains.**

12. You have heard so much about the importance of reducing fat consumption. Why then are Jeff's allowable fat units so high? **At his advanced level of physical fitness, Jeff will use up the energy from the fat before the fat can cause harm in his body.**

Copyright © by the Glencoe Division of Macmillan/McGraw-Hill School Publishing Company

160

CHAPTER 21 ASSESSMENT
TRANSPORT

Understanding Concepts

In the space at the left, write the letter of the word or phrase that best completes the statement or answers the question.

c 1. The transport system of vascular plants consists of
a. roots and stems.   c. xylem and phloem.
b. buds and leaves.   d. cells and root hairs.

a 2. The transport of oxygen in the blood is the function of the
a. red blood cells.   b. white blood cells.   c. antibodies.   d. antigens.

c 3. Cell fragments that help repair cuts by clotting the blood are
a. red blood cells.   b. white blood cells.   c. platelets.   d. antigens.

Write the word or phrase that best completes the statement.

4. After **xylem** cells in plants die, they form hollow tubes that transport water to the leaves and other parts of the plant.

5. The process by which water is lost through the pores in stomata is called **transpiration**.

6. Oxygen, carbon dioxide, and nitrogenous wastes pass through the capillary wall by the process of **diffusion**.

7. The largest blood vessel in the body is the **aorta**.

In the space at the left, write the letter of the phrase from Column B that best matches the term in Column A.

	Column A	Column B
g	8. root hair	a. proteins on the membranes of red blood cells
d	9. vein	b. place where an electrical impulse causes the ventricles to contract
f	10. lymph node	c. site where gases and nutrients are exchanged between the body cells and the blood
i	11. white blood cell	d. carries blood to the heart
a	12. antigens	e. living phloem cells that transport sucrose
e	13. sieve tube	f. place where most bacteria in the body are destroyed
h	14. artery	g. absorbs water and dissolved ions from the film of moisture around soil particles
c	15. capillary	h. carries blood away from the heart
j	16. type O blood	i. cell that defends against viruses, microorganisms, and parasites
b	17. atrioventricular node	j. universal donor

TRANSPORT

Interpreting and Applying Concepts

To demonstrate how sucrose, a simple sugar, is transported in a plant, researchers girdled a tree. Girdling involves removing a ring of bark around the tree. The girdle cuts through and destroys the phloem tissue. It has no effect on the xylem, which is closer to the center of the tree stem. Study Figure 21-1 and answer the following questions.

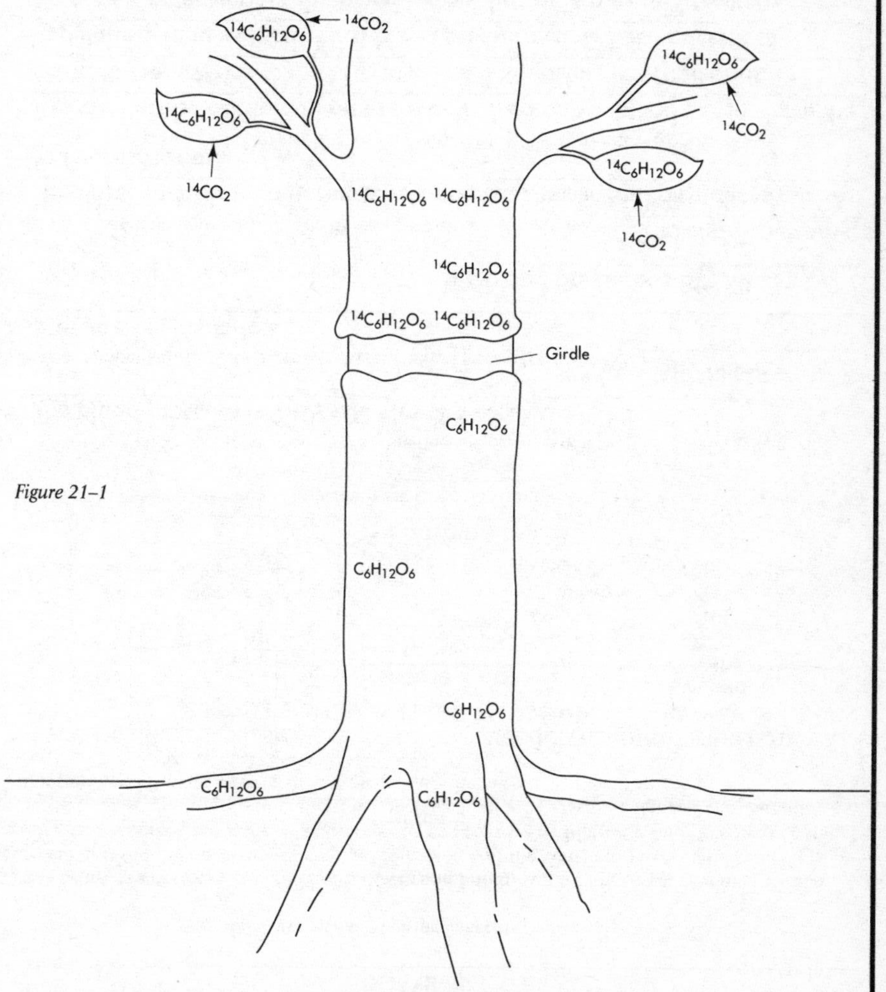

Figure 21-1

161

TRANSPORT

Interpreting and Applying Concepts   continued

Blood pressure changes as the blood flows through the body. A blood pressure reading includes the *systolic pressure,* the pressure during the contraction of the ventricles, over the *diastolic pressure,* the pressure during the relaxation of the ventricles. The highest blood pressure occurs in the ventricles. Use the information about the changes in blood pressure (Figure 21–2) and the velocities of blood flow to answer the following questions.

VELOCITIES OF BLOOD FLOW

Vessels	Velocity (cm/sec)
Aorta	40
Arteries	40–10
Arterioles	10–0.1
Capillaries	0.1
Venules	0.3
Veins	0.3–5.0
Vena Cavae	5–20

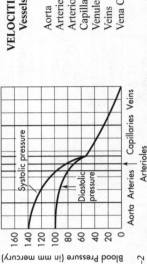

Figure 21–2

7. What is the patient's blood pressure as shown in the graph when the blood enters the aorta? **The blood pressure is 140/100 mm of mercury.**

8. What happens to the blood pressure as the blood enters the arteries? **The blood pressure begins to fall.**

9. What is the blood pressure in the arterioles? **The blood pressure is 80/70 mm of mercury.**

10. What change do you notice in the blood pressure of the capillaries? **The capillaries have no pulse. There is no difference between the systolic and diastolic pressures in the capillaries.**

11. Relate the pressures to the velocities of the blood as it flows through the body. **Blood velocities and blood pressure both drop as the blood passes from the aorta to the arteries to the arterioles, until it reaches the capillaries. In capillaries, venules, and veins, blood pressure continues to decrease. Velocities increase after the blood leaves the capillaries and continue to do so until the blood returns to the heart.**

12. Why is it a selective advantage that the blood slows down in the capillaries? **The slow rate of flow in the capillaries allows time for the exchange of nutrients and gases between the blood and the tissue cells.**

Copyright © by the Glencoe Division of Macmillan/McGraw-Hill School Publishing Company

---

TRANSPORT

Interpreting and Applying Concepts   continued

1. The researchers supplied carbon dioxide labeled with radioactive carbon (carbon–14) to the leaves of the girdled tree. What happened to the radioactive carbon in the leaves? **It was used to make simple sugars $^{14}C_6H_{12}O_6$ in the leaves.**

2. Where is the radioactive carbon found in the tree stem? **The radioactive carbon is found above the girdle but not below it.**

3. Study Figure 21–2 to see if the radioactive carbon reached the roots. **There is no radioactive carbon in the roots.**

4. What can you deduce about the role of phloem in the transport of sucrose from the leaves? Why is there a bulge in the stem above the girdle? **The phloem transports sucrose from the leaves through the stem to where the girdle destroyed the phloem. Above the girdle there is a bulge caused by the radioactive sugars that welled up when they could not continue to move below the level where the bark was damaged.**

5. How could the experimenters be certain that xylem tubes had not transported the radioactive sugars? **If the xylem were the transporter of sucrose, there would have been radioactive sucrose in the lower tree stem and in the root. The xylem was not destroyed by the girdling.**

6. How do you know that the sucrose in the lower stem and roots was produced in the leaves before the girdling took place? **The sucrose in the lower stem and leaves had to have been produced before the girdling occurred because it did not contain radioactive carbon. The researchers controlled the carbon dioxide supplied to the leaves after the girdling. Only carbon dioxide with radioactive carbon was supplied after the girdling, so all the sucrose made after that had to be radioactive, too.**

Copyright © by the Glencoe Division of Macmillan/McGraw-Hill School Publishing Company

GAS EXCHANGE AND EXCRETION

Interpreting and Applying Concepts

At elevations of about 3500 meters above sea level, the decreased amount of oxygen available begins to affect the human respiratory system. People experience the symptoms of altitude sickness, including dizziness and blurred vision. The graphs in Figure 22–1 show the results of some studies conducted to determine the effects of high altitudes on humans. Study the graphs and answer the following questions in complete sentences.

Figure 22–1

A. Effect of Breathing Air or Oxygen on Saturation of Oxygen in Arteries

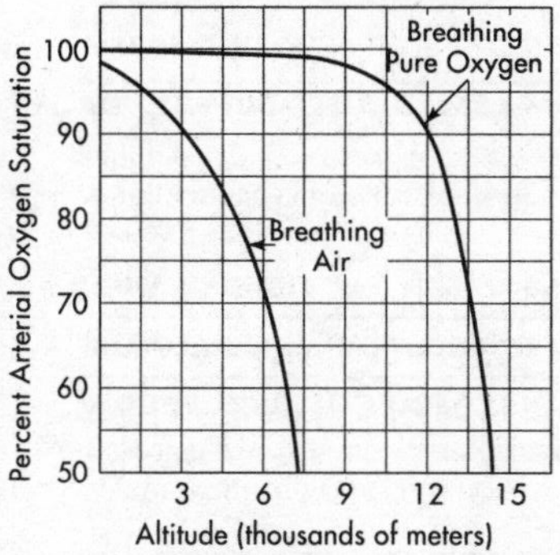

B. Effect of Altitude on Time Before Beginning of Collapse and Coma

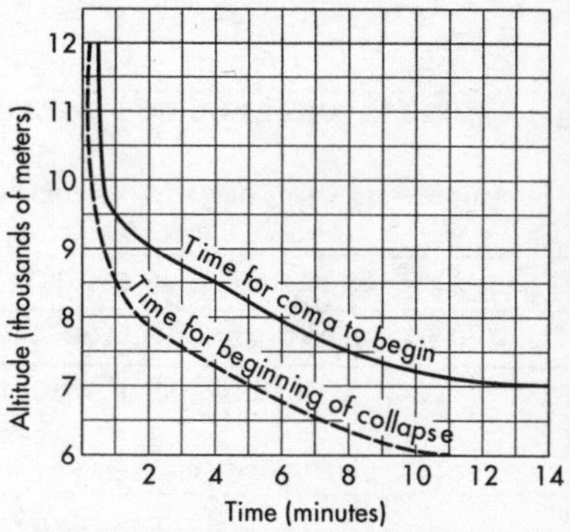

1. Refer to Figure 22–1A. Up to what altitude does the oxygen in the arteries of a person breathing pure oxygen remain close to 100 percent saturation? How does this compare to someone breathing air? **Up to around 9000 m the oxygen saturation in the arteries of a person breathing pure oxygen remains close to 100%. The oxygen saturation in the arteries of a person breathing air would be considerably lower than 50% at 9000 m.**

2. Refer to Figure 22–1A. The lowest level of arterial oxygen at which a person can remain alive for more than a few hours is 50 percent. What is the highest altitude at which someone breathing pure oxygen can stay alive? **At about 14 000 m the oxygen in the arteries is reduced to 50 % when someone is breathing pure oxygen.**

3. Refer to Figure 22–1A. Some people experience mild altitude sickness when breathing air at 3000 m. At what saturation level is the oxygen in their arteries? **The percent of oxygen saturation is 90 % at 3000 meters.**

4. Refer to Figure 22–1A. There is an altitude above which a person cannot survive breathing air. There is a different altitude above which a person cannot survive breathing pure oxygen. How do these two altitudes compare? **The altitude above which a person cannot survive breathing pure oxygen is twice the altitude above which a person cannot survive breathing air.**

Copyright © by the Glencoe Division of Macmillan/McGraw-Hill School Publishing Company

---

CHAPTER 22 ASSESSMENT

GAS EXCHANGE AND EXCRETION

Understanding Concepts

In the space at the left, write the letter of the word or phrase that best completes the statement or answers the question.

c   1. The kangaroo rat does not drink water because
- a. water is harmful to its body.
- b. it does not need water to live.
- c. its kidneys resorb almost all the water that passes through them.
- d. its food is mostly liquid.

a   2. The excess salt taken in by marine fish is removed by
- a. active transport across the gills.
- b. the kidneys.
- c. being absorbed by the cells.
- d. nephridia.

d   3. A paramecium maintains osmotic balance by removing excess water by means of
- a. cilia.
- b. excretory ducts.
- c. flame cells.
- d. contractile vacuoles.

b   4. In humans, most of the water flowing through the kidneys
- a. stays in the glomerulus.
- b. is resorbed by the kidneys.
- c. passes out in the urine.
- d. remains in the stomach.

c   5. In fish, oxygen-rich water
- a. leaves through the mouth.
- b. is expelled through the gills.
- c. passes over the gills, where oxygen is removed.
- d. is transported through the body.

In the space at the left, write the letter of the phrase from Column B that best matches the term in Column A.

	Column A	Column B
i	6. vasopressin	a. tube through which urine is excreted
d	7. ureter	b. muscular storage sac for urine
f	8. glomerulus	c. removal of nitrogenous wastes
b	9. urinary bladder	d. tube from the kidney that conveys urine to the bladder
e	10. flame cell	e. part of the excretory system in a planarian
a	11. urethra	f. mass of capillaries in a nephron
h	12. Malpighian tubule	g. excretory organ in an earthworm
j	13. nephron	h. excretory organ in a grasshopper
c	14. excretion	i. hormone that controls the amount of water excreted
g	15. nephridium	j. tiny excretory unit in the kidney

GAS EXCHANGE AND EXCRETION

Interpreting and Applying Concepts continued

5. Refer to Figure 22–1B. The great danger of high altitudes is the impairment of a person's ability to respond when deprived of sufficient oxygen. "Collapse" means extreme weakness and mental haziness. "Coma" means unconsciousness. Find out how long a mountain climber has to replace an oxygen mask ripped off during a fall while almost at the peak of Mount Everest (8900 m). **The mountain climber has slightly less than one minute before showing signs of collapse and about 2 1/2 minutes before becoming unconscious.**

6. A person who stays at a high altitude for several weeks usually will recover from altitude sickness. The body adapts in the following ways: the bone marrow produces greater numbers of red blood cells than usual, the respiratory center of the brain is stimulated, and the size of the alveoli may increase. Explain how each of these changes helps the person overcome the effects of the high-altitude conditions. **The increased production of red blood cells means the ability of the blood to transport oxygen is also increased; stimulation of the brain's respiratory center causes deeper and more rapid breathing; larger alveoli can take in more air to help compensate for the reduced amount of oxygen at the higher altitude.**

A water environment contains certain ions (mostly sodium, Na^+, chlorine, Cl^-, and magnesium, Mg^{++}). The ions combine to form salts. Often the concentrations of these ions in the bodies of the animals that live in water differ from the concentrations of ions in the water in which they live. If an animal's ionic concentrations are higher than its surroundings, it tends to gain water and lose ions. If an animal's ionic concentrations are lower than its surroundings, it loses water and gains ions. Study the graph in Figure 22–2, which compares the relative concentrations of ions in animals that live in water with the concentrations of the ions in seawater and in fresh water. Then answer the following questions in complete sentences.

Figure 22-2

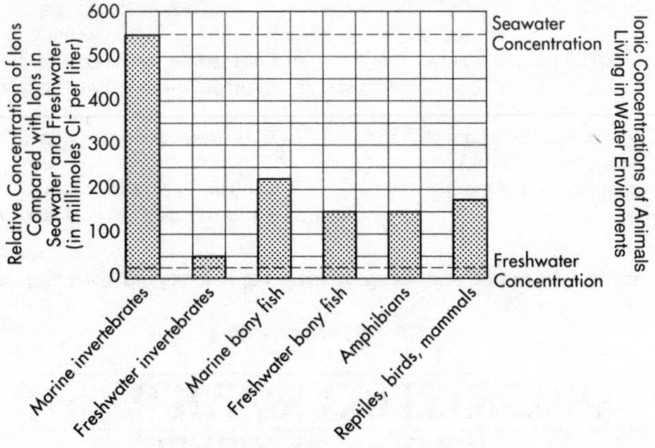

Ionic Concentrations of Animals Living in Water Enviroments

GAS EXCHANGE AND EXCRETION

Interpreting and Applying Concepts continued

7. Which animals have body fluids with an ionic content closest to that of seawater? **Marine invertebrates have the same ionic concentrations as seawater.**

8. Which animals have ionic concentrations similar to that of fresh water? **Freshwater invertebrates have similar ionic concentrations to those in fresh water.**

9. How might you explain why the ionic concentrations of marine bony fish are closer to that of fresh water than to the seawater in which it lives? **Marine bony fish descended from freshwater fish, so it is reasonable that they would have body fluids somewhat similar to those of freshwater fish.**

10. Compare the ion concentrations in freshwater bony fish with that of their environment. What must these fish do to survive in their environment? **Since the ionic concentration of these fish is greater than that of their environment, they would gain water and lose ions. Their adaptations include not drinking water and not losing any more ions than occurs through diffusion.**

11. Compare the ion concentrations in marine bony fish with that of their environment. How can these fish survive in their environment? **Marine bony fish have a less concentrated ion content than is in their environment. They tend to lose water to their environment and to take in ions. To overcome this, they drink large quantities of seawater to get rid of excess salt.**

12. Salmon move between freshwater and seawater environments. How must they change their habits when they move to a new environment? **As they move from fresh water to sea water, salmon must take in more water and get rid of excess ions.**

CHAPTER 23 ✤ ASSESSMENT
IMMUNE SYSTEM

Understanding Concepts

In the space at the left, write the letter of the word or phrase that best completes the statement or answers the question.

c 1. The bacteria that cause strep throat are
   a. endotoxins. **b.** exotoxins. **c.** pathogens. **d.** macrophages.

a 2. A specific defense mechanism that guards your body from disease-causing microorganisms is
   a. release of histamines. **c.** saliva.
   b. hairs in the nasal passage. **d.** unbroken skin.

a 3. White blood cells that are produced in the bone marrow but mature in the thymus gland are
   a. T cells. **b.** B cells. **c.** receptors. **d.** antigens.

d 4. Which of the following are not part of the inflammatory response?
   a. fever
   b. swelling
   c. production of large numbers of white blood cells
   d. acids and enzymes in your body

Write the word or phrase that best completes the statement.

5. Poisonous chemicals secreted by living bacteria are known as __exotoxins__.

6. __Lymphocytes__ are white blood cells that recognize antigens and begin to destroy specific pathogens.

7. A(n) __cytotoxic T cell__ defends against body cells in which viruses are reproducing.

8. A(n) __autoimmune disease__ is an abnormal response of the immune system to part of a person's own body.

9. A person who receives antibodies made by another animal acquires a type of protection against disease called __passive immunity__.

10. __Receptors__ are proteins in T cells that recognize certain self proteins.

11. The lymph vessels, lymph nodes, spleen, tonsils, bone marrow, and thymus gland are all part of the __immune system__.

12. __Macrophages__ are white blood cells that engulf and digest pathogens.

Copyright © by the Glencoe Division of Macmillan/McGraw-Hill School Publishing Company

---

IMMUNE SYSTEM

Interpreting and Applying Concepts

Inbred mice in which brothers and sisters are mated can produce offspring that are homozygous for a certain trait. Continued crossing for about 20 generations can produce mice in which all the traits are exactly the same. Two different types of purebred mice—A and B—were produced in this way. Then A mice were crossed with B mice; the offspring were called A × B mice. The parents involved in the cross were then given radiation treatment to destroy all their lymphocytes. The irradiated parent mice immediately received donations of lymphocytes from the bone marrow or spleen of their A × B offspring. The lymphocytes from an offspring's bone marrow were immature at the time of donation, that is, they had not yet learned to distinguish between self and non-self proteins. Lymphocytes from spleen cells were already mature when donated; they had learned to distinguish between self and non-self before donation. The lymphocyte donations caused the A and B mice to have lymphocytes that were exactly like the donor's, while the mice's other tissues were as they had been before the donation. Table 23–1 shows what happened when the irradiated mice with donor lymphocytes were vaccinated with A cells infected with vaccinia virus (A-vaccinia) or B cells infected with vaccinia virus (B-vaccinia). The vaccinations were carried out for a total of four experiments, and the results were recorded in the table. (Note: The vaccinia virus causes cowpox. This is the same virus used by Edward Jenner almost 200 years ago to produce the first vaccinations against smallpox.) Using the information in Table 23–1, answer the following questions.

Table 23–1

Experiment	Donor Cells	Recipient	Response to infected cells	
			A-vaccinia	B-vaccinia
1	A × B (from bone marrow)	A	+	–
2	A × B (from bone marrow)	B	–	+
3	A × B (from spleen)	A	+	+
4	A × B (from spleen)	B	+	+

"+" indicates an immune response; "–" indicates no response

1. In Experiment 1, lymphocytes from both types A and B bone-marrow cells were placed in irradiated type A mice. What happened when either vaccinia-A or vaccinia-B was given to the mice? **The donated bone marrow lymphocytes produced an immune response to vaccinia-A, but not to vaccinia-B.**

2. In Experiment 2, both types of bone-marrow lymphocytes were placed in irradiated type B mice. What was the result when either vaccine was administered? **The donated bone marrow lymphocytes produced an immune response to vaccinia-B, but not to vaccinia-A.**

3. In Experiment 3, both types of lymphocytes that had matured in spleen cells were placed in irradiated type A mice. What was the result when either vaccine was administered? **The lymphocytes from spleen cells placed in type A mice produced an immune response to both kinds of vaccines.**

4. In Experiment 4, both types of lymphocytes that matured in spleen cells were donated to irradiated type B mice. Describe their reaction to both kinds of vaccines. **The lymphocytes from spleen cells placed in type B mice produced an immune response to both kinds of vaccines.**

Copyright © by the Glencoe Division of Macmillan/McGraw-Hill School Publishing Company

165

IMMUNE SYSTEM

Interpreting and Applying Concepts   continued

5. How would you explain the results of these experiments? **The immature A and B bone marrow lymphocytes matured in the mice to which they were donated. The lymphocytes learned to recognize as an antigen the vaccine of the same type as the mice into which it was injected. The mature A and B lymphocytes from the spleen had already learned to react to both kinds of antigens, no matter in which type of mice they were placed.**

In other experiments, mice of type C normally reject grafts from type D mice. However, newborn mice of type C can be given cells from type D mice. If six weeks later, type C mice receive a skin graft from mice of type D and E, they accept the grafts from type D mice and reject the grafts from type E mice.

6. Why did the newborn type C mice accept cells from type D mice? **The lymphocytes of newborn mice had not yet learned to recognize what is self and what is non-self.**

7. Why did the six-week old type C mice accept the grafts from type D mice but not the grafts from type E mice? **The type C mice accepted the grafts from type D mice because they had received cells from type D mice before they learned to recognize self from non-self proteins. They accepted type D cells as self from that time on. The grafts from type E mice were rejected because the type C mice had not learned to accept them as self.**

You have learned that T cells recognize certain self proteins. These proteins are called HLA, which stands for "human leucocyte antigens." Unless you are an identical twin, you have a combination of different HLA on the surface of your cells that is not likely to be duplicated in anyone else. The reason for the variation in HLA is not known, but it has been suggested that by carrying a large number of different HLA, your immune system is less likely to be invaded by a microbe that imitates one of your self proteins. Table 23-2 lists some diseases associated with certain HLA. The relative risk column shows the extent to which an individual carrying a particular HLA is more likely to contract the disease when compared to a person without the antigen. Study the table. Then answer the questions that follow.

Table 23-2

Disease	Antigen	Relative Risk
Ankylosing spondylitis	B27	87.8
Rheumatoid arthritis	DRw4	4.0
	DR4	6.0
	B8	1.0
Multiple sclerosis	A3	1.8
	B7	2.0
	Bw2	1.9
	DRw2	3.8

IMMUNE SYSTEM

Interpreting and Applying Concepts   continued

Disease	Antigen	Relative Risk
Addison's disease	Dw3	8.8
	DR3	6.5
	B8	4.0
Grave's disease	B8	2.5
	Bw35	5.0
	Dw3	5.5
Juvenile diabetes	B8	2.7
	DR3	5.0
	DR4	5.0
	DR3 and DR4	14.0

8. How are a person's HLA a defense against disease? **The HLA recognize self and non-self proteins in the body. This alerts the immune system that an unrecognized protein is present and must be acted upon.**

9. What disease is associated with HLA B27? What risk is there that a person with B27 will more likely contract the disease than a person who lacks the antigen? **The B27 antigen is associated with ankylosing spondylitis. The risk that a person who has the antigen will have the disease is 87.8 times the risk for a person who lacks the antigen.**

10. How much greater is the risk of developing juvenile diabetes for an individual who has both DR3 and DR4 than for someone who has only one of these antigens? **An individual with both DR3 and DR4 has 2.8 times greater risk of developing juvenile diabetes than someone with only one of these antigens.**

11. To which diseases is a person with Dw3 susceptible? **Someone with Dw3 is susceptible to Addison's disease and Grave's disease.**

12. In a population, every individual has his or her own, different HLA combination. How might this difference in self-proteins help preserve the species? **Since each individual has different self-proteins, it is unlikely that a pathogen would evolve with the ability to avoid the immune response in all the individuals in an entire population.**

13. If, as self proteins, the HLA help protect the body, hypothesize why the HLA listed in Table 23-2 are linked with diseases that destroy the body. **Any reasonable hypothesis is acceptable. Answers may include that some change causes a breakdown in the ability to recognize certain antigens as self or a change in certain self proteins may make them no longer recognizable as self. The body is then compelled to destroy these antigens.**

CHAPTER 24 ❧ ASSESSMENT
CHEMICAL CONTROL

Understanding Concepts

Write the word or phrase that best completes the statement.

1. Any plant response caused by unequal stimulation is called a(n) **tropism**.

2. **Auxins** are plant growth hormones produced in the tips of stems.

3. **Apical dominance** occurs when hormones from the tip of a stem inhibit the development of lateral buds.

4. A group of plant hormones called **inhibitors** promote dormancy in seeds and buds.

5. In animals, chemical control is regulated by hormones secreted by the **endocrine glands**.

In the space at the left, write the letter of the phrase from Column B that best matches the term in Column A.

	Column A		Column B
i	6. gibberellins	a.	growth response of a plant to gravity
j	7. phototropism	b.	hormone that triggers the uptake of glucose from the blood stream
f	8. protein-type hormones	c.	hormones that cause cell elongation and apical dominance in plants
d	9. pituitary gland	d.	gland that controls the secretion of hormones from many other endocrine glands
a	10. geotropism	e.	the dropping of fruit or leaves from a plant
b	11. insulin	f.	cause a response in cells through the action of a second messenger
e	12. abscission	g.	a brain structure that exerts control over the pituitary gland
g	13. hypothalamus	h.	hormones that activate genes within cells of target tissue
h	14. steroid hormones	i.	growth hormones that cause elongation, increase germination rate and fruit production of some plants
k	15. FSH	j.	growth response of a plant to light
m	16. pancreas	k.	released by the pituitary gland early in the menstrual cycle
n	17. vasopressin	l.	hormone that regulates metabolic rate of body cells
l	18. thyroxine	m.	gland that plays a vital role in regulating blood glucose levels
c	19. auxins	n.	hormone that helps maintain the body's water balance

---

CHEMICAL CONTROL

Interpreting and Applying Concepts

The drawing in Figure 24-1 represents a plant shoot. Study Figure 24-1 and answer the questions that follow.

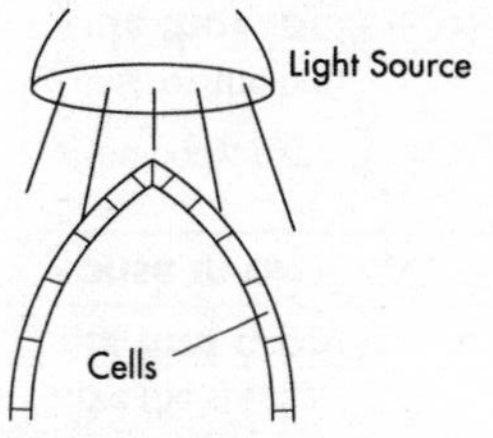

Light Source

Cells

Figure 24-1

1. In the space provided, sketch the same shoot as it would appear after several days of growth. Be sure to include individual cells in your drawing.

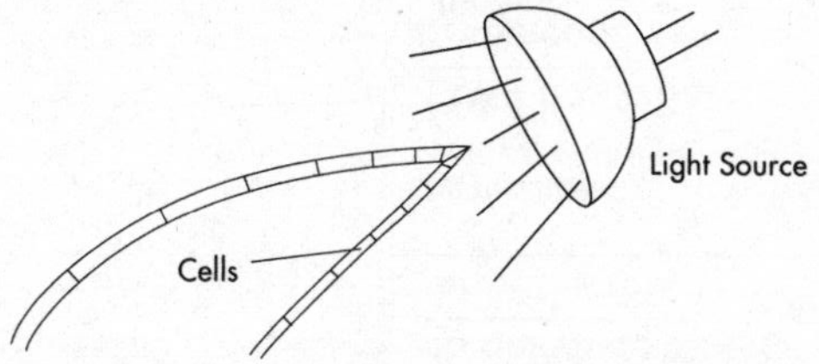

Light Source

Cells

2. How would your sketch have been different if the light source had been to the left of the shoot? Directly above the shoot? **If the light source had been to the left of the shoot, the cells on the right side would be elongated and the plant tip would bend toward the left. If the light source had been directly above the shoot, the cells would elongate equally and the shoot would not be bent.**

Answer the following questions in complete sentences.

3. A landscaper wanted to have bushy shrubs in front of a house and conical shrubs along the driveway leading to the house. How could this combination of shapes be achieved using the same kind of plant throughout? **Select a plant with apical dominance, which will naturally produce cone-shaped shrubs. The shrubs in front of the house can be made bushy by pinching or cutting off the tips of the plants as they grow. The shrubs along the driveway need no alteration.**

4. Why do you think the addition of gibberellins makes many dwarf plants grow tall while it has little effect on plants that are naturally tall or grow as vines? **The plants that grow naturally tall probably produce their own gibberellin. Addition of more has no effect.**

167

CHEMICAL CONTROL

Interpreting and Applying Concepts continued

5. What effect might the application of gibberellins to grass seed planted in early spring have on the thickness of the lawn produced from this seed? **Since gibberellins increase the rate of germination of many types of seed, you might expect a thicker, healthier lawn from the treated seed.**

6. As part of a negative feedback loop to maintain a constant level of calcium in the blood, the thyroid gland releases a hormone called calcitonin whenever the level of calcium in the blood is raised. Predict what happens next with regard to the calcium level in the blood and the secretion of calcitonin. **The calcitonin signals a body response that lowers the level of calcium in the blood and this, in turn, causes the thyroid to stop secreting calcitonin.**

7. The pituitary is often referred to as the "master gland." Explain why this is an appropriate name for this gland. **Many of the hormones secreted by the pituitary are directed to other endocrine glands around the body. These pituitary hormones direct the activities of these other glands.**

8. A human embryo produces progesterone. What effect does this action have on the menstrual cycle of a pregnant woman? Explain the reason for this effect. **Progesterone inhibits the production of FSH and LH. These two hormones are needed during the early stages of the menstrual cycle. Thus, the production of progesterone causes menstrual cycles to cease during pregnancy.**

9. Hypoglycemia is a condition in which the glucose level in the blood falls to extremely low levels. Describe two ways that this condition might be treated. **Eating foods that contain sugar, such as candy, can raise the blood glucose level quickly. The injection of glucagon into the bloodstream causes the liver to break down glycogen and release glucose into the bloodstream.**

10. Both human growth hormone (HGH) and thyroxine are needed for normal growth. What are the effects of HGH and thyroxine? How might a lack of thyroxine inhibit a person's growth? **HTH stimulates increase in size only. Thyroxine increases the metabolic rate of body cells as they grow. A lack of thyroxine slows cell activity and thus inhibits cell enlargement and division, resulting in slow body growth.**

11. In a person with untreated diabetes mellitus, would you expect secretion of glucagon to be increased or decreased? Explain. **People with untreated diabetes mellitus have high levels of blood glucose. The effect of glucagon is to increase blood glucose levels. Thus, glucagon secretion would be decreased in people with diabetes mellitus.**

Copyright © by the Glencoe Division of Macmillan/McGraw-Hill School Publishing Company

168

CHAPTER 25 ❧ ASSESSMENT
NERVOUS CONTROL

NERVOUS CONTROL

Understanding Concepts

In the space at the left, write the letter of the word or phrase that best completes the statement or answers the question.

c   1. Which is *not* required in order for a nervous response to occur?
     a. stimulus detection   c. hormone secretion
     b. effector response   d. impulse transmission

a   2. In most animals, incoming impulses are received and transmitted by
     a. sensory neurons.   c. interneurons.
     b. motor neurons.   d. response neurons.

d   3. In a nerve cell, impulses are transmitted away from the cell body by
     a. effectors.   b. dendrites.   c. myelin.   d. an axon.

c   4. All of these can affect the speed with which nerve impulses are conducted *except*
     a. myelin sheath.   b. nodes.   c. polarity.   d. axon diameter.

b   5. Chemicals involved in moving nerve impulses across a synapse are called
     a. neuroexciters.   b. neurotransmitters.   c. receptor molecules.   d. neuroinhibitors.

b   6. Chemical substances that reduce nerve transmission across the synapse are called
     a. stimulants.   b. depressants.   c. neurotoxins.   d. inhibitors.

d   7. The simplest phylum of animals that possess a nervous system is
     a. monera.   b. porifera.   c. planaria.   d. cnidaria.

b   8. The largest part of the human brain is the
     a. medulla.   b. cerebrum.   c. brainstem.   d. cerebellum.

a   9. Sympathetic and parasympathetic nerves are controlled by the
     a. medulla oblongata.   c. cerebral cortex.
     b. cerebellum.   d. thalamus.

c   10. The part of the human brain most closely associated with endocrine control is the
     a. cerebral cortex.   b. pons.   c. hypothalamus.   d. midbrain.

c   11. Involuntary or automatic responses to stimuli
     a. do not involve the CNS.   c. are called reflexes.
     b. are regulated by hormones.   d. are controlled by the autonomic system.

b   12. Touch is a general term that includes each of these related senses *except*
     a. pain.   b. brightness.   c. cold.   d. pressure.

NERVOUS CONTROL

Interpreting and Applying Concepts

Use the letters of the steps listed below to answer the question that follows. Write the letters in the spaces provided.

   a. impulses cross synapses and activate muscle cell effectors

   b. impulses cross synapses and enter spinal cord interneurons

   c. muscles contract, moving the hand

   d. impulses interpreted as heat and pain

   e. some impulses cross synapses to motor neurons, which leave the spinal cord

   f. sensory neurons are stimulated

   g. receptors in the skin are activated

   h. some impulses travel to the brain

   i. impulses travel to the spinal cord

1. If your hand touches a hot stove, the reflex response occurs in two stages—initial and secondary. In what order do the steps occur?

initial: $\underline{g} \rightarrow \underline{f} \rightarrow \underline{i} \rightarrow \underline{b} \rightarrow \underline{e} \rightarrow \underline{a} \rightarrow \underline{c}$

secondary: $\underline{h} \rightarrow \underline{d}$

Answer the following questions in complete sentences.

2. The human tongue can detect four different taste sensations—sweet, salt, acid, and bitter. These sensations are detected by sensors called taste buds. A particular type of taste bud is specific for one taste sensation only. For example, taste buds that are stimulated by sweet substances are not stimulated by sour, salty, or bitter substances. Different types of taste buds are concentrated in different regions of the tongue.
     Describe an experiment that could help you "map" the locations of these different taste regions on the tongue.

Student responses may vary, but might include using dilute solutions of salty, sweet, sour, and bitter substances to test their effects on different parts of the tongue. Emphasis should be placed on using substances that can be safely ingested.

NERVOUS CONTROL

Interpreting and Applying Concepts   continued

3. Why is it important that reflex actions be involuntary rather than voluntary? **Many reflex actions are responses to stimuli that present a danger to the organism. These responses must take place very quickly. Since involuntary actions do not involve the brain, they can take place almost immediately after receptor cells are stimulated.**

4. Predict the initial and secondary reflex responses to hitting your right thumb with a hammer. **The initial response would be the jerking back of the right hand; the secondary responses would be the sensation of pain in the right thumb and verbal expression of that pain.**

5. Acetylcholine causes muscles to contract and also to relax. Explain this seeming contradiction. **The effect of acetylcholine on skeletal muscle is different from its effect on heart muscle. In response to a stimulus, acetylcholine is the neurotransmitter that carries nerve impulses across the synapses and causes the contraction of skeletal muscles. In the autonomic system, the action of acetylcholine is the reverse. The sympathetic nervous system speeds up heart and lung action during stress by stimulating heart muscle contractions. After the stressful situation is over, the parasympathetic nervous system stimulates the release of acetylcholine, which inhibits heart muscle contractions.**

6. Some poisons, such as curare, bind to the receptors for the neurotransmitter acetylcholine on muscles, inhibiting these muscles. What effect does this action have on the human body? **Acetylcholine excites the neurons that activate skeletal muscle. A poison that interferes with the function of acetylcholine causes paralysis of skeletal muscles, including those involved in breathing.**

7. People who have lost an arm or leg due to accident or surgical amputation often report feeling sensations at the site of the missing body part. What is a likely explanation for this phenomenon? **The endings of nerve fibers that once extended into the missing part are exposed in the scar tissue at the amputation site. Stimulation of these nerve endings may be interpreted by the brain as originating at the original site of the nerve endings.**

Copyright © by the Glencoe Division of Macmillan/McGraw-Hill School Publishing Company

170

CHAPTER 26 ASSESSMENT

MOVEMENT

Understanding Concepts

Answer the following questions.

1. What material makes up most of the skeleton of the fetus of a vertebrate? **cartilage**

2. What structural characteristic is unique to vertebrates? **endoskeleton**

3. Why do arthropods molt? **to grow, or increase in size**

4. Name three important functions of an exoskeleton. **protection, support, movement, and prevention of water loss (for terrestrial arthropods)**

5. What are the individual parts that make up an endoskeleton? **bones**

6. What is the process by which cartilage is replaced by bone? **ossification**

7. What are two nutrients needed for proper bone development? **calcium compounds and vitamin D**

8. Where are Haversian canals found and what is their function? **in compact bone; to provide channels for vessels and nerves**

9. What are the connective tissues of a moveable joint? **ligaments**

10. Name the four types of moveable joints in vertebrates. **ball-and-socket, hinge, pivot, and gliding joints**

In the space at the left, write TRUE if the statement is true. If the statement is false, change the italicized word or phrase to make it true.

skull	11. In a human skeleton, fixed joints are found in the *wrist and ankle*.
true	12. When a person's triceps muscle contracts, his or her arm *straightens*.
striated	13. Muscles that move the skeleton are made up of *cardiac* muscle tissue.
arthropods	14. In addition to vertebrates, *cnidarians* make up the only other animal group to produce movement through the combined actions of muscles and skeletons.
true	15. The sliding filament hypothesis explains the process of *muscle contraction*.
involuntary	16. Nerves of the autonomic system control the action of *voluntary* muscles.
true	17. The human hip is an example of a *ball-and-socket* joint.
true	18. Bones are formed by special cells called *osteoblasts*.

MOVEMENT

Interpreting and Applying Concepts

Figure 26–1 shows a working model constructed by a student to show how opposing muscles produce motion around a joint. The model consists of two pieces of wood joined together with elastic tape, a colored balloon, a white balloon, and some velcro strips used to attach the balloons to the pieces of wood. Study Figure 26–1 and answer the questions that follow.

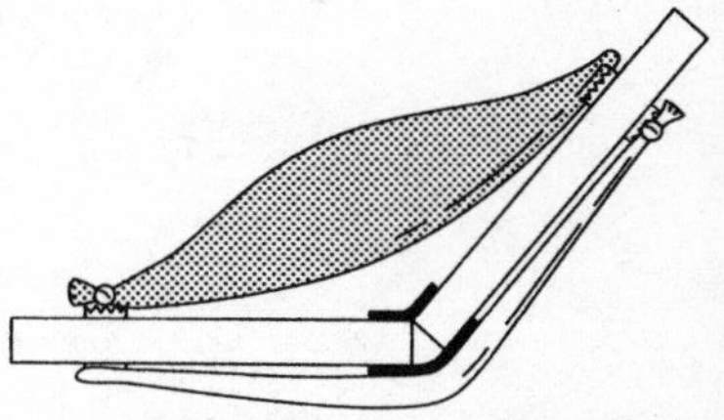

Figure 26–1

1. What do the pieces of wood represent? **bones**

2. What kind of joint is represented? **hinge**

3. What does the elastic tape represent? **ligaments**

4. What do the velcro strips represent? **tendons**

5. What kind of muscle is represented by the colored balloon? The white balloon? **flexor; extensor**

6. Where are similar joints and muscle pairs found in the human body? Name the joints. **arm-elbow; leg-knee**

7. What effect would detaching one of the velcro strips from the white balloon have on the ability of the muscle pair to perform their tasks? **The flexor could still cause the joint to bend, but the extensor could no longer straighten out the joint.**

8. Suppose you were asked to design an experiment to test the hypothesis that muscle fatigue affects coordination. Describe your experiment. **Answers will vary. Students may describe recording the time it takes for subjects to perform some task requiring hand-eye coordination, such as threading a needle. As a control, have the subjects perform the task prior to any exercise. Then have them repeat this task at various intervals, each time after performing vigorous exercise involving the hands and/or arms, such as push-ups or squeezing a hard rubber ball.**

171

MOVEMENT

Interpreting and Applying Concepts   continued

Answer the following questions in complete sentences.

9. Although the skeletal system consists largely of hard, bony tissue, it also contains a considerable amount of softer, more flexible tissues. Name two types of soft tissue found in the skeletal system and describe the function of each.

Cartilage, which is found at the ends of bones and in the disks between vertebrae, provides flexibility and cushions against impact or pressure. Ligaments hold bones together at joints.

10. Most vertebrate organisms are weakest and most vulnerable to injury or capture by predators when they are very young. They become less vulnerable as they grow larger and stronger. Arthropods, on the other hand, remain vulnerable at times well into adulthood. Explain.

Arthropods have exoskeletons. In order to grow, they must molt, or shed these exoskeletons. During the interval between molting and hardening of their new exoskeletons, arthropods have little protection from enemies or predators.

11. Describe some methods of movement used by organisms that may lack skeletons or muscles or both. **Answers may vary but may include the following: Organisms that have cilia and/or flagella utilize energy from ATP to move these structures. This motion propels the organisms through water. Jellyfish move by contracting their muscles, which forces water from under the medusa and produces a jet-propulsion effect, moving jellyfish through the water. Methods of locomotion of earthworms, clams, and starfish may also be described.**

12. In humans, the Achilles tendon, located at the lower back portion of the leg, attaches the calf muscles to the heel bone of the foot. Because of its location, the Achilles tendon is vulnerable to injury, especially during athletic events. How would severing the Achilles tendon affect body movement? **The calf muscles are flexors. When they contract, they pull up on the back of the foot, causing the toes of the foot to point down. Thus, a person with a severed Achilles tendon would be unable to walk or even to bend his or her foot up and down.**

13. How might the movement of the human body be affected if striated muscle tissues were replaced with smooth muscle tissues over which one had voluntary control? **Since smooth muscle contracts and relaxes more slowly than skeletal muscle, rapid voluntary movements would be impossible.**

POPULATION BIOLOGY

Interpreting and Applying Concepts

Each of the graphs in Figure 27–1 shows a relationship between change in population size and time. Refer to the graphs as you answer the following questions.

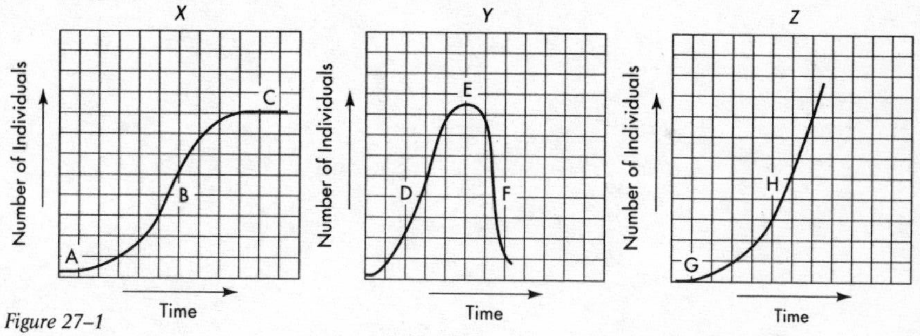

Figure 27–1

1. Which graph represents growth under ideal conditions? How do you know?
Graph Z; it indicates that population growth continues indefinitely.

2. Describe what is happening in graph Y to the population size at point F.
It is decreasing rapidly.

3. In graph X, what is happening to the population size at points A, B, and C?
A-increasing slowly; B-increasing most rapidly; C-remaining steady.

4. How does the birthrate of the organisms compare with the death rate at point C? At point F?
C-birthrate and death rate are equal; F-death rate greatly exceeds birthrate

5. If each graph were extended beyond the time limits shown, which would show a continued growth in population size?
graph Z

6. The curve of which graph most closely resembles that expected for a frog population in a pond? For a lightning bug population?
frog population-graph X; lightning bug population-graph Y

7. Which graph shows the effects of density-dependent limiting factors on population size? What might some of those limiting factors be?
Graph X; density-dependent factors may include disease, predators, lack of food, and overcrowding.

Copyright © by the Glencoe Division of Macmillan/McGraw-Hill School Publishing Company

---

CHAPTER 27 ❧ ASSESSMENT
POPULATION BIOLOGY

Understanding Concepts

In the space at the left, write the letter of the phrase from Column B that best matches the term in Column A.

Column A

b   1. social hierarchy
d   2. biotic potential
i   3. population density
f   4. interspecific competition
j   5. intraspecific competition
a   6. limiting factors
h   7. carrying capacity
e   8. population growth curve
c   9. emigration
g   10. predation

Column B

a. circumstances that prevent organisms from reaching their biotic potential
b. a designated chain of command based on dominance
c. the moving of a population out of an area
d. the highest rate of reproduction under ideal conditions
e. a population graph having an S shape
f. competition among populations of different species
g. the feeding of one organism on another
h. the maximum number of individuals in a given population that the environment can support
i. the size of a population that occupies a given area at any given point in time
j. competition between members of the same species

Write the word or phrase that best completes the statement.

11. **Zero population growth** is a condition in which the birthrate equals the death rate and the population is not growing in size.

12. Social hierarchy is also known as a **pecking order**.

13. **Extinction** occurs when a species dies out due to competition with a different species for the same resources.

14. Food supply and living space are examples of **density-dependent** limiting factors.

15. The swarming of bees from an overcrowded hive to a new location is an example of **emigration**.

Copyright © by the Glencoe Division of Macmillan/McGraw-Hill School Publishing Company

POPULATION BIOLOGY

Interpreting and Applying Concepts   continued

Answer the following questions in complete sentences.

8. Both interspecific competition and intraspecific competition are density-dependent limiting factors. Why is extinction a possible outcome of interspecific competition, but not of intraspecific competition?

Interspecific competition takes place among populations of different species. If one species dominates another, the weaker species may be killed off and become extinct. Intraspecific competition takes place among members of the same species. Dominant members of a given species may destroy weaker members, but the species itself remains in existence.

Questions 9 and 10 are related questions. Where appropriate, use the terms *limiting factors, carrying capacity,* and *population density* in your answer.

9. Suppose you were to place several mating pairs of rabbits on a small island having ample vegetation and no predators. Describe the changes you would expect to see in the rabbit population over time. Include a discussion of any limiting factors in your description.

Students should describe a steady increase in population size until the population density reaches the carrying capacity of the island. Once the carrying capacity is reached, population size should remain fairly stable. Students should identify food supply as being the most obvious limiting factor. Overcrowding and disease might also be suggested as limiting factors.

10. Once the island's carrying capacity for rabbits is reached, several mating pairs of foxes are introduced to the island. What effect will this have on the rabbit population of the island? Describe the changes in rabbit and fox populations you would expect to take place over time.

The introduction of foxes, natural predators of rabbits, will cause a decrease in the rabbit population. As long as an ample food supply (rabbits) for the foxes exists, the fox population will continue to grow. However, as the rabbit population declines, the fox population will begin to decrease due to a limited food supply and perhaps also to overcrowding and disease. As this happens, the rabbit population will begin to increase once again.

POPULATION BIOLOGY

Interpreting and Applying Concepts   continued

The age structure diagrams in Figure 27–2 show how individuals are distributed at each age level for different human populations. Study the diagrams and then answer the questions.

Figure 27-2

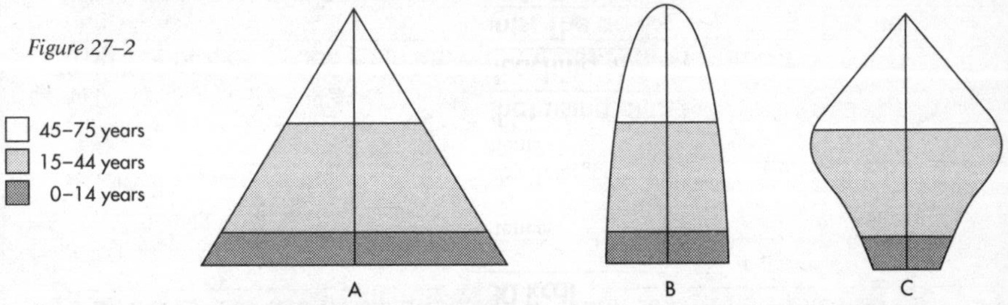

- 45–75 years
- 15–44 years
- 0–14 years

A   B   C

11. Based on what you know about average reproductive age for humans, which graph do you think represents a rapidly expanding population? Explain.

Graph A represents a rapidly expanding population. It shows a large reproductive-age population plus a large number of children who will move into that category in the next 15 years.

12. Describe the types of populations represented by the two graphs you did not name in your answer to Question 11.

Graph B represents a slowly expanding population; Graph C represents a declining population.

CHAPTER 28 ❦ ASSESSMENT

ECOSYSTEMS

Understanding Concepts

In the space at the left, write the letter of the word or phrase that best completes the statement or answers the question.

d   1. All the possible feeding levels within an ecosystem make up
     a. a food chain.   b. a biome.   c. abiotic factors.   d. a food web.

b   2. A feeding step within an ecosystem is also known as
     a. a pyramid.   b. a trophic level.   c. a niche.   d. a producer.

b   3. Relationships between or among different organisms in an environment are known as
     a. interactive biomass.   c. diversifications.
     b. biotic factors.   d. communal activities.

a   4. A rabbit that eats a plant is
     a. a first-order consumer.   c. a second-order consumer.
     b. a producer.   d. a carnivore.

d   5. All first-order consumers are
     a. carnivores.   b. omnivores.   c. small animals.   d. herbivores.

d   6. The pyramid of numbers does not apply to food chains in which
     a. consumers are at the lowest trophic level.
     b. carnivores outnumber herbivores.
     c. producers are at the higher trophic levels.
     d. small organisms feed on a large organism.

b   7. What type of relationship between two organisms results in benefits to both organisms?
     a. parasitism   b. mutualism   c. equalism   d. commensalism

c   8. All of the interacting biotic and abiotic factors in an environment make up
     a. a food chain.   b. a population.   c. an ecosystem.   d. a biomass pyramid.

Answer the following questions.

9. What term is used to identify an organism's place in an ecosystem?   **niche**

10. What happens to most of the potential energy that is "lost" between links of a food chain? **It is changed to heat energy during cellular respiration.**

11. What is the term used to identify organisms that break down and consume organic materials and wastes in a food chain? **decomposers**

12. What is the trophic level of a person eating a piece of bread?   **first-order consumer**

ECOSYSTEMS

Interpreting and Applying Concepts

Answer the following questions.

1. In the space provided, construct a food web made up of the following organisms: hawk, field mouse, deer, grass, grass, shrubs, cricket, rabbit, frog, mountain lion. Then describe the food chains in your food web.

Food webs may vary somewhat, but should include the following food chains: grasses and shrubs, rabbit, mountain lion; grasses and shrubs, rabbit, hawk; grasses, field mouse, hawk; grasses, field mouse, snake, hawk; grasses, field mouse, owl, hawk; grasses, cricket, frog, owl, hawk; grasses and shrubs, deer, mountain lion.

2. Consider a hypothetical situation in which 20 000 kilocalories of energy is available to the producers in a food chain. First-order consumers receive 20 percent of this energy, and the available energy is reduced by half for each of the next two levels. Determine the amount of energy available at the first, second, and third trophic levels of this food chain.

First level: 20 000 kcal × 0.20 = 4000 kcal;
second level: 4000 kcal × 0.50 = 2000 kcal;
third level: 2000 kcal × 0.50 = 1000 kcal

Answer the following questions in complete sentences.

3. Many conservation-minded people use laundry detergents that do not contain phosphates. How does such a practice help the environment?
Student responses should indicate that using phosphate-free detergents can reduce the problem of phosphates leaching into the waters of lakes and ponds. Phosphates are plant nutrients. The accumulation of these nutrients in a lake or pond promotes increased algal growth. When algae exhaust the nutrient supply, they die. Decomposition uses up oxygen dissolved in the water. Other organisms die and more oxygen is removed. Eventually such bodies of water become unfit for most living things.

ECOSYSTEMS

Interpreting and Applying Concepts   continued

4. Compare and contrast commensalism, mutualism, and parasitism. **All three involve the relationship between a host and another organism. In commensalism, the host organism is neither aided nor harmed, while the other organism benefits from the relationship. In mutualism, both the organisms benefit from the relationships. In parasitism, the host organism is harmed by the action of the other organism, which benefits from the relationship.**

5. Why are fuels such as coal, natural gas, and petroleum products called fossil fuels? Trace the energy stored in these fuels back to its original source—energy from the sun. **These fuels are so-named because they are formed from the remains of once-living things. All living things derive their energy from the sun. Green plants use the sun's energy directly to make food. This energy is passed on to consumers. All organisms store some of this energy in their cells. This is the energy that is stored in fossil fuels.**

6. How would a food web in a desert area be similar to one in a tropical rain forest? How would it be different? **Food webs in both environments would contain producers, consumers, and decomposers, and the sun would be the source of energy in both. A food web in a tropical rain forest may consist of many more and varied organisms than one in a desert.**

7. Only a small percentage of the sun's energy that reaches Earth's surface is converted to chemical energy through photosynthesis. What happens to most of the sun's energy that reaches Earth and how does this affect Earth's environments? **Most of the sun's energy is absorbed by Earth's surface and converted to heat. This heat then raises the temperature of Earth and its atmosphere.**

8. Why is the diversity of life greater near the shorelines and near the water's surface than in deeper waters? **Responses may vary, but should indicate the effects of decreasing temperature and light with increasing depth. Students may also mention the fact that near the shorelines, the water is shallow enough to allow plant roots to reach the soil beneath the water and to allow plant leaves to remain above water.**

ECOSYSTEMS

Interpreting and Applying Concepts   continued

9. Loam is a soil consisting of a mixture of clay, silt, sand, and a small amount of gravel. A topsoil consisting of loam mixed with humus is more fertile than topsoils consisting of humus mixed with clay or with sand alone. Explain. **Because of the varying particle sizes in loam, it has a better capacity for holding water than does sand, but it does not prevent water from percolating, as does clay. Plant roots can move easily through loamy soils and derive nutrients dissolved in the water as it moves slowly through these soils.**

10. Instead of leaving land bare, farmers often plant legumes, such as clover or alfalfa, on land that is not being used to grow cash crops. Why? **The plants help to prevent erosion of topsoil by wind and water, and the nitrogen-fixing bacteria that live in the roots of the legumes convert nitrogen to ammonia and other nitrogen compounds that can be used as nutrients by these plants and can enrich the soil as well.**

ORIGIN AND DISTRIBUTION OF COMMUNITIES

Understanding Concepts

In the space at the left, write the letter of the biome from the following list that best matches each description. A biome may be used more than once.

a. desert   b. grassland   c. taiga   d. temperate forest   e. tropical rain forest   f. tundra

c   1. coniferous trees

f   2. permanently frozen soil

b   3. prairie, steppe, pampa, veldt

a   4. little rainfall; rapid evaporation

d   5. deciduous trees

f   6. low temperature; short summer

b   7. uneven rainfall; scattered trees

c   8. fog; water-soaked, acidic soil

d   9. woodland with definite seasons

e   10. constant temperature; heavy rainfall

In the space at the left, write the letter of the word or phrase that best completes the statement or answers the question.

a   11. Two abiotic factors affecting the distribution of organisms in the ocean are
    a. light and temperature.
    b. humidity and altitude.
    c. temperature and precipitation.
    d. light and latitude.

c   12. Each of the following is characteristic of a climax community *except*
    a. large plants with long life cycles.
    b. steady population sizes.
    c. simple food chains.
    d. diversity of species.

c   13. The top layer of vegetation in a temperate or tropical rain forest is the
    a. biosphere.   b. benthos.   c. canopy.   d. shrub layer.

d   14. The circulation of water in the ocean is mostly affected by water temperature and the
    a. movement of ocean animals.
    b. tilt of Earth's axis.
    c. topography.
    d. wind.

b   15. A possible sequence of plants in primary land succession is
    a. grasses , shrubs, mosses, trees, ferns.
    b. lichens, mosses, grasses, shrubs, trees.
    c. trees, mosses, lichens, shrubs, grasses.
    d. shrubs, grasses, ferns, mosses, lichens.

In the space at the left, write **TRUE** if the statement is true. If the statement is false, change the italicized word or phrase to make it true.

permafrost   16. The *biosphere* is a layer of soil that never thaws during the tundra summer.

true   17. Desert communities tend to form on the *leeward* side of mountains.

Nekton   18. *Plankton* move freely through the ocean under their own power.

ORIGIN AND DISTRIBUTION OF COMMUNITIES

Interpreting and Applying Concepts

A climatogram is a graph that displays average monthly temperature and precipitation for a particular region. The months of the year are given along the bottom. The bar graph indicates average monthly precipitation and uses the scale along the left side. The line graph indicates average monthly temperature and uses the scale along the right side.

The table below gives the average annual precipitation and temperature range for the five biomes you studied. Use the table and climatograms in Figure 29-1 to answer the questions.

Biome	Average Annual Precipitation (cm)	Range of Average Monthly Temperatures (°C)
Tundra	less than 25	−24° to 4°
Temperate Forest	75 to 100	5° to 29°
Tropical Rain Forest	more than 200	24° to 28°
Grassland	25 to 75	0° to 24°
Desert	less than 25	25° to 33°

Figure 29–1

ORIGIN AND DISTRIBUTION OF COMMUNITIES

Interpreting and Applying Concepts   continued

1. What is the range of the average monthly temperatures of the region shown in climatogram A?
about 1°C to 26°C

2. What is the total annual precipitation of the region shown in climatogram C?
about 16–17 cm

3. Based on precipitation, what two biomes could climatogram C represent?
tundra or desert

4. What other factor(s) could you use to determine which biome climatogram C represents?
range of average monthly temperatures

5. What type of biome does climatogram B represent? Explain.
tundra; monthly temperature ranges from −16°C to 10°C with low average annual precipitation.

6. Name some animals that are likely to inhabit the region shown in climatogram B.
Answers may vary; reindeer, caribou, wolves, snowshoe hares.

7. Which climatogram represents a tropical rain forest biome? Explain your reasoning.
D; monthly average temperature is about 25°C; total annual precipitation is greater than 200 cm

8. How might a climatogram representing a grassland biome compare with climatogram A?
Range of monthly temperatures of grassland is similar to that of climatogram A, but may be slightly lower. Total annual precipitation is less in grassland biome and would be heaviest during the summer months.

Copyright © by the Glencoe Division of Macmillan/McGraw-Hill School Publishing Company

---

ORIGIN AND DISTRIBUTION OF COMMUNITIES

Interpreting and Applying Concepts   continued

The hot, sandy beach area along a lake or ocean shore may be marked by sand dunes built up by the action of the wind. The four illustrations in Figure 29–2 show the primary succession in such a sand dune community, but the stages are not shown in order. Study Figure 29-2, then answer the questions.

Figure 29-2

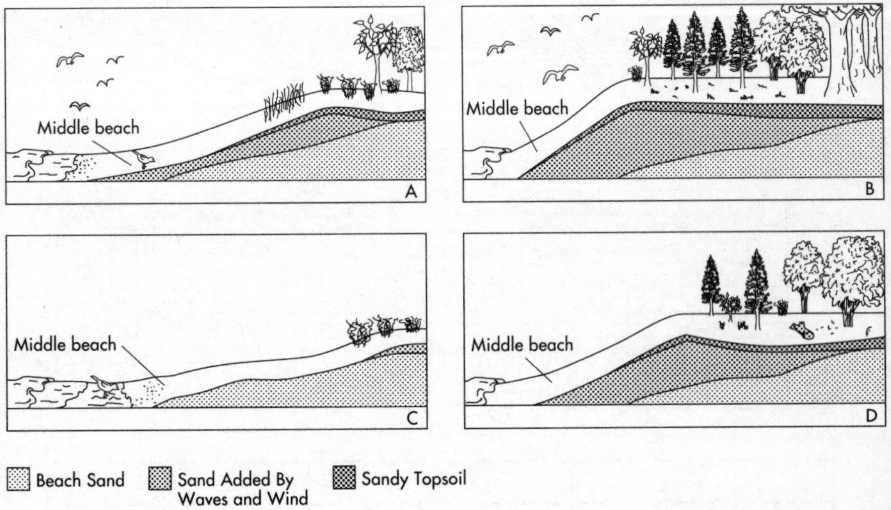

Beach Sand | Sand Added By Waves and Wind | Sandy Topsoil

9. Sequence the stages of succession from pioneer stage to climax community. **C, A, D, B**

10. What factors helped you identify the pioneer stage? **few small simple plants**

11. How did you identify the climax stage? **large complex plants of many species**

12. What types of animals are likely to inhabit the area during the stage shown in illustration A?
Answers may vary; small rodents, insects, some birds

13. What factors might explain why no vegetation develops in the middle beach area throughout the stages of succession?
Wind and wave action keep shifting and moving the sand so plant roots can't anchor.

14. How do the pioneer plants in the sand dune succession differ from the pioneer plants in the forest succession shown on page 816 of your textbook? **There are no rocks for lichens and mosses to grow on. Pioneer plants are grasses and small shrubs, which have extensive root systems, so they are better able to anchor plants in the sand.**

Copyright © by the Glencoe Division of Macmillan/McGraw-Hill School Publishing Company

CHAPTER 30 ❧ ASSESSMENT

HUMANS AND THE ENVIRONMENT

Understanding Concepts

In the space at the left, write TRUE if the statement is true. If the statement is false, change the italicized word or phrase to make it true.

true 1. *Habitat destruction* is a major cause of the extinction of plant and animal species.

erosion 2. The removal of soil by the action of moving air and water is *pollution*.

nonbiodegradable 3. Compounds that cannot be broken down into inactive chemicals by living organisms are *nonrenewable*.

pesticides 4. *Chlorofluorocarbons* are chemicals used to kill unwanted organisms.

depleted 5. Soil that is no longer fertile enough to support crop growth because its nutrient content has been reduced is said to be *eroded*.

true 6. Integrated pest management includes the breeding of pest-resistant plants and the *use of pesticides*.

In the space at the left, write the letter of the possible solution or alternative from the following list that matches the given environmental problem. A problem may have more than one solution, and a solution may be used more than once.

a. alternative energy sources b. autocidal controls c. ban CFC use d. biological controls
e. coal scrubbers f. contour plowing g. crop rotation h. reforestation i. terracing

a, e 7. acid precipitation

c 8. depletion of ozone

f, h, i 9. erosion

a, h 10. greenhouse effect

b, d 11. pesticide use

a, e 12. photochemical smog

g 13. soil depletion

In the space at the left, write the letter of the word or phrase that best completes the statement or answers the question.

b 14. Each of the following is an alternative to pesticide use *except*
  a. autocidal control.   c. biological control.
  b. biological magnification. d. cultural control.

d 15. Food and forests are considered
  a. nonrenewable resources.   c. biodegradable.
  b. nonbiodegradable.   d. renewable resources.

a 16. The alternate planting of soil-enriching and soil-depleting crops on an area of land is
  a. crop rotation.   c. terracing.
  b. strip cropping. d. the greenhouse effect.

HUMANS AND THE ENVIRONMENT

Interpreting and Applying Concepts

Pesticide X is a nonbiodegradable pesticide that was used to treat plants in a forest community. Study the chart and food web shown in Figure 30–1, then answer the questions.

Representative Concentrations of Pesticide X in Tissues (parts per million, ppm)

Producers	0.01–0.05
First-order consumers	0.25–1.5
Second-order consumers	2.1–4.5
Third-order consumers	4.1–13.8
Fourth-order and above	4.5 +

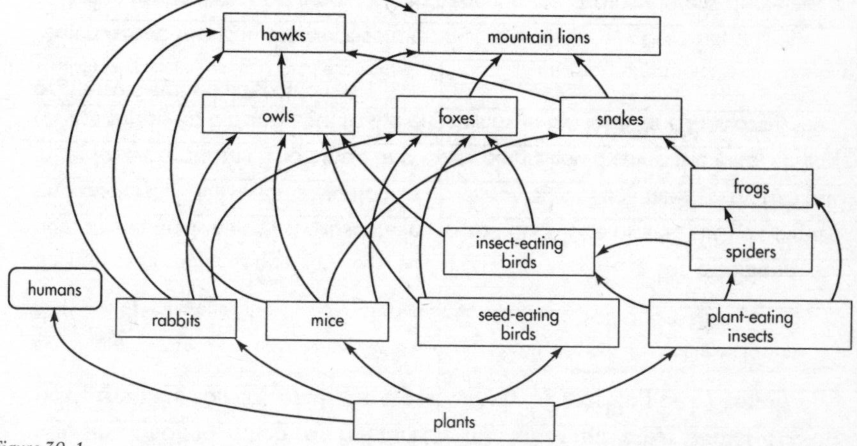

Figure 30–1

1. About how many times higher is the average concentration of pesticide X in primary consumers than in producers? **about 25 to 30 times**

2. Which animal(s) in the food web will have the highest concentrations of pesticide X in their bodies? **owls, foxes, snakes, hawks, and mountain lions**

HUMANS AND THE ENVIRONMENT

Interpreting and Applying Concepts   continued

3. Would a snake that primarily feeds on frogs tend to have a higher or lower concentration of pesticide X in its tissues than a fox that primarily feeds on rabbits? Explain your reasoning. **Answers will vary. The snake would have a higher concentration of pesticide in its tissues because it is a fourth-order consumer in its food chain with the frogs; the fox would be a second-order consumer in its food chain with the rabbits.**

4. If humans ate rabbits, snakes, and frogs, about how much pesticide X (in parts per million) would you expect to find in the tissues of a human that was part of this food web? **Answers will vary. Anywhere from 2.1 to 13.8 ppm.**

Complete Table 30-1 and answer the questions.

Table 30-1

Pollutant	Sources	Negative Effect on Environment
carbon monoxide	fossil fuel burning	oxygen deficiency in humans
carbon dioxide	fossil fuel burning	greenhouse effect
sulfur dioxide	fossil fuel burning	acid precipitation
nitrogen oxides	fossil fuel burning	tropospheric ozone formation, acid precipitation, smog
CFCs	aerosol sprays	stratospheric ozone depletion
pesticides	agricultural uses	biological magnification
hydrocarbons	fossil fuel burning	photochemical smog
biological wastes	industry, agriculture, humans	oxygen deprivation in lakes and ponds

5. What is the major contributor to the pollution of the environment? **the burning of fossil fuels**

HUMANS AND THE ENVIRONMENT

Interpreting and Applying Concepts   continued

6. What part of the environment (air, water, soil) is most affected by carbon dioxide? What part is most affected by pesticides? **air; soil**

7. How do sulfur dioxide and nitrogen oxides pollute our waters? **When these substances combine with water they form acids. These acids reach Earth in the form of acid precipitation which can fall directly into lakes and other bodies of water. Acids from the soil can enter the water supply by leaching or by runoff.**

8. Fertilizers, which are used to add nutrients to depleted soil, can become pollutants. What part(s) of the environment would be affected by fertilizer use? **soil and water**

9. If fertilizers run off into lakes, how could they affect a lake ecosystem? **The additional nutrients in the lake could cause the overgrowth of lake plants, such as algae, and bacteria. The bacteria would use up dissolved oxygen needed by fish and other lake organisms. The algae and other plants could cover the lake, cutting off the supply of air and light to the organisms in the lake. In either case, the organisms in the lake could die.**

10. Explain why ozone formation and ozone depletion both have negative effects on the environment. **When ozone forms in the troposphere, the layer of air we breathe, it is irritating to the eyes and lungs. When ozone in the stratosphere is depleted, a greater amount of harmful ultraviolet radiation can reach the surface of Earth.**

180